Outdoor Inquiries

Outdoor Inquiries

Taking Science Investigations Outside the Classroom

Patricia McGlashan, Kristen Gasser,
Peter Dow, David Hartney, and Bill Rogers

Foreword by Hubert M. Dyasi

HEINEMANN
Portsmouth, NH

Heinemann
A division of Reed Elsevier Inc.
361 Hanover Street
Portsmouth, NH 03801–3912
www.heinemann.com

Offices and agents throughout the world

The authors and publisher wish to thank those who have generously given permission to reprint borrowed material:

Excerpt from *The Trees in My Forest* by Bernd Heinrich. Copyright © 1997 by Bernd Heinrich. Reprinted by permission of HarperCollins Publishers.

Figure 9–1: from *Inquiry and the National Science Education Standards: A Guide for Teaching and Learning.* Copyright © 2000 by the National Academy of Sciences. Reprinted with permission of the National Academies Press, Washington, D.C.

 This material is based upon work supported by the National Science Foundation under Grant No. 9911354.

Any opinions, findings, and conclusions or recommendations expressed in this material are those of the author(s) and do not necessarily reflect the views of the National Science Foundation.

About the cover photographs: Mirror images of the same outdoor inquiry study site were photographed 20 days apart in an investigation of eight such rectangles along a city street. See details of the study at www.firsthandlearning.org and link to enewsletter in Archives.

Library of Congress Cataloging-in-Publication Data
Outdoor inquiries : taking science investigations outside the classroom / Patricia McGlashan . . . [et al.] ; foreword by Hubert M. Dyasi.
     p. cm.
   Includes bibliographical references.
   ISBN-13: 978-0-325-01120-2
   ISBN-10: 0-325-01120-6
   1. Nature study—United States.   2. Nature study—Activity programs—United States.
I. McGlashan, Patricia.

LB1585.3.O88 2007
372.35′7—dc22   2007000625

Editor: Robin Najar
Production: Vicki Kasabian
Cover design: Jenny Jensen Greenleaf
Typesetter: House of Equations, Inc.
Manufacturing: Steve Bernier

Printed in the United States of America on acid-free paper
11 10 09 08 07 EB 1 2 3 4 5

Contents

Acknowledgments

Writing this book has been a collective enterprise. It represents the combined efforts and insights of a team of scientists, science teachers, and curriculum specialists who have been working together for many years on the development of natural history–based science experiences for elementary and middle-grade students. These efforts have been guided by a National Advisory Board that included Mark Baldwin of the Roger Tory Peterson Institute, Deanna Banks Beane of the Association of Science-Technology Centers, Hubert Dyasi of the City College of New York, Myles Gordon of the American Museum of Natural History, Lynn Rankin of the Exploratorium, Karen Worth of Wheelock College, and Bernie Zubrowski of Education Development Center.

Several teachers and scientists have provided illustrative vignettes from their work. These include Mark Baldwin; Wayne Gall, Regional Entomologist for the New York State Department of Health; Gail Grigg, teacher in the Royalton-Hartland Central School District; Allan Hayes, teacher at the Nichols School; Richard S. Laub, Curator of Geology at the Buffalo Museum of Science; Michael Milliman, teacher in the Amherst School District; Sara R. Morris, Professor of Biology at Canisius College; Elizabeth S. Peña, archaeologist and Director of the Art Conservation Department at Buffalo State College; and Bill Rogers, Director of Out-of-School Programs at First Hand Learning. Other Buffalo-area educators who have made valuable contributions to the text include Paula Connors, Barbara Shaughnessy, Michelle Rose, and Janet Siulc. In addition, Buffalo Zoo President Donna Fernandes and Curator of Education Tiffany Vanderwerf helped shape the ideas for the study of animal behavior that appear in Chapter 8.

The idea for this book grew out of First Hand Learning's commitment to helping teachers make use of the physical surroundings of the school as a catalyst for scientific inquiry. Staff contributors include Peter Dow, David Hartney, Julian Montague, Benjamin O'Brien, and Bill Rogers. The final design is the work of Patricia McGlashan, who served as lead writer, and Kristen Gasser, who was the managing editor. Financial support was provided by The John R. Oishei Foundation and the National Science Foundation (Award ID #9911354). Any opinions, findings and conclusions, or recommendations expressed in this material are those of the authors and do not necessarily reflect the views of the funders.

First Hand Learning, Inc.
2495 Main Street, Suite 559
Buffalo, NY 14214
www.firsthandlearning.org

First Hand Learning Staff

Delores Anderson, Operations Assistant
Michelle DeFrancesco, Manager of Operations
Peter Dow, President
Kristen Gasser, Director of Publications
David Hartney, Managing Director
Patricia McGlashan, Director of Curriculum Development
Julian Montague, Art Director
Benjamin O'Brien, Assistant Art Director
Bill Rogers, Director of Out-of-School Programs
Richard Rotolo, Science Materials Center Manager
Marilyn Sozanski, Project Assistant

Outdoor Inquiries

Introduction

Students at all grade levels and in every domain of science should have the opportunity to use scientific inquiry and develop the ability to think and act in ways associated with inquiry, including asking questions, planning and conducting investigations, using appropriate tools and techniques to gather data, thinking critically and logically about relationships between evidence and explanations, constructing and analyzing alternative explanations, and communicating scientific arguments.

National Research Council
National Science Education Standards

Imagine the following scene: it is spring and students in a metropolitan community are escaping their classrooms by going outside to study the various organisms that inhabit their schoolyards, neighborhoods, and local parks. Each child has a field journal for making drawings, recording observations, and writing down questions that arise from these firsthand investigations. At one location the teacher gives out plastic storage bags to her students, who have collected a variety of treasures including soil samples, leaves, seeds, and other intriguing objects. At another school the children have hand lenses that they are using to examine tiny plants growing in sidewalk cracks. Suddenly they spy a parade of tiny ants making its way from the sidewalk to the base of a tree. "Why are the ants going up the tree?" one writes, while another student continues to make a list of all the organisms they have found. At a third school, outside the city, yet another group of budding naturalists is vigorously manning homemade sweep nets in an open grassy area where they are capturing flying insects, including grasshoppers, froghoppers, and even a praying mantis, while others are placing tiny spiders in collection jars for further study. At a fourth location we find a class working with a compass, long measuring tape, and a collection of wooden stakes, laying out a grid in the field behind their school. Soon they will be plotting the locations of the trees and pathways in their schoolyard on a large sheet of construction paper.

Returning to their classrooms, the teachers of these students review and comment on the range of drawings, descriptions, and questions that appear in the field journals. They select several examples and ask the class to comment on them, reinforcing the power of observational diversity and praising each student's efforts. Visitors to these classrooms would see that these children are engaged in building their own unique knowledge base about the natural and built environments that surround them. They would find this knowledge reflected in the journals, in the objects that they had collected for further study, in their efforts to create field guides to their local habitats, and in the rich discussions that the children were having about their experiences and about organisms and objects they have found. What might impress the visitors most is the depth of student engagement in their personal discoveries and in the range of their questions. And while struck by their factual knowledge, they might well recall Rachel Carson's caveat about the importance of cultivating sensory experience in the young: "If facts are the seeds that later produce knowledge and wisdom, then the emotions and the impression of the senses are the fertile soil in which the seeds must grow. The years of childhood are the time to prepare the soil."[1]

We hope that this vision of schoolchildren engaged in rich, diverse, and self-motivated investigations will become a reality for teachers pursuing *Outdoor Inquiries*. According to the *National Science Education Standards* (National Research Council 1996), and many of the state standards as well, all students should be able to *"think and act in ways associated with inquiry."* *Outdoor Inquiries* seeks to meet that challenge by giving teachers and students a framework for carrying out their own firsthand investigations in the schoolyard and in their neighborhoods. Doing scientific inquiry in this real-world context involves the use of specific strategies that focus on gathering data, keeping good records, and making careful observations that may take place over a long period of time. E. O. Wilson, in his autobiography, *Naturalist*, discusses the special character of field-based natural science:

> The experiments are conducted in opposite manner from those in conventional laboratories. They are retroactive rather than anticipatory. Whereas most biologists vary a few factors under controlled conditions and observe the effects of each deviation, the evolutionary biologist observes the results already obtained, as learned from studies of natural history, and tries to infer the factors that operated in the past. Where the experimental biologist predicts the outcome of experiments, the evolutionary biologist retrodicts the experiment already performed by Nature; he teases science out of history.[2]

In other words, doing scientific investigation the way a field naturalist does involves close observations, long-term data gathering, and the search for questions rather than laboratory-based, controlled experimentation driven by anticipated answers.

What can students learn about scientific inquiry from careful, disciplined observations of the familiar world of plants and animals that we encounter every day? Think about a class of elementary-aged children looking at a group of squirrels chasing each other in the maple tree outside their classroom window. It is early September and the tree is laden with seeds that are beginning to fall to the ground. From the shape of the leaves and seeds they could easily identify it from a field guide as a Norway maple, but to know much more about this particular tree, its life cycle, and its relationship to the other organisms in its habitat, they would have to watch it regularly and keep track of their observations. They might notice, later in the fall when most of its leaves had dried up and fallen, that a squirrel was gathering up some of them and returning them to the tree. Later, when the branches are totally bare, they might see clusters of those leaves located at the extremities of several branches. Are they nests? How could they find out? By now they have no doubt observed several squirrels chasing each other through those bare branches. Did these squirrels use the gathered leaves to build those structures? Is this where they protect and rear their young? If so, which squirrels are the parents, when did they mate, and how do they care for their offspring? What do they eat, and how do they produce enough food and warmth to survive a harsh, northern winter? Perhaps they have seen a squirrel perched on a branch chewing nuts on a sunny day in January, but where did the nut come from? And, incidentally, what is that twitchy tail flicking all about? Our young observers will have these and many more questions about their tree and its squirrel inhabitants, but to answer them they will need to observe, record, and ask questions in order to make sense out of all their random bits of information. In short, they must structure their inquiry.

In this book we have assembled a set of strategies that teachers and others can use to help young people become inquiring naturalists. In a world inundated with prepackaged information of all kinds, *Outdoor Inquiries* is dedicated to helping young people develop their knowledge of the world around them by gathering their own data and analyzing it for themselves. To help them accomplish this we have outlined five interrelated strategies: journal keeping, mapping, field guide development, collection making, and behavior study. None of these strategies is particularly unique, but taken together they provide a collective way of assembling a database of information that can be used by successive groups of students to build an understanding of their immediate surroundings. By recording the natural history of a specific location, such as a schoolyard or neighborhood park, they will give present and future occupants of this particular habitat a storehouse of information that they can use for continuing investigation and research. In this way they can become experts in the study of their own backyard. To illustrate the value of this approach, we have included brief personal accounts of how both teachers and scientists have used this sort of data gathering in their own search for knowledge and understanding.

The primary audience for *Outdoor Inquiries* is teachers and mentors of upper-elementary and middle school students, both in formal and informal settings. Others will find the approach useful as well, but we have chosen this age level because we have found that it is a time when students are particularly prone to act like scientists. They are beginning to question what they are being taught, and they are growing in their ability to think logically and to reason from evidence. They are also still quite curious about how the world works. Helping them to channel that curiosity into critical thinking and logical reasoning is our central objective. *Outdoor Inquiries* is an outgrowth of a kit-based program called *Object Lessons* that uses authentic objects as a starting point for scientific investigation of natural and cultural phenomena. In "Examining Mammals," for example, students investigate five "mystery skulls" to determine the differences between carnivores and herbivores and then, by pursuing further clues provided by fur, feet, tracks, and scat, discover the identity of each of the animals. In "Investigating Insects" and "Digging Archaeology" students employ similar object-based strategies to solve mysteries using firsthand evidence. One of the benefits of this evidence-based approach to understanding is that it encourages close observation, drawing, vocabulary development, and both descriptive and analytical writing. When children are solving a mystery, or pursuing answers to their own questions, they become highly motivated to think critically, and to develop their communication skills. Because we found this firsthand approach to learning to be so popular with teachers and students, we have extended it to the more complex and less predictable environment beyond the classroom door. In the outdoor classroom students can now chose what to study, but we have provided a set of strategies and guidelines for organizing that study.

The field journal provides the foundation for the learning process proposed in this book. It is the place where students make their drawings, record their observations and questions, and keep a running record of their fieldwork. It also provides a window into student thinking for the teacher that can be valuable in guiding their investigations. Mapmaking provides a tool for recording the locations of important objects and significant events, for *where* things happen is often as important as *when* things happen in natural history–based investigations. The field guide is a document that brings together the individual findings and journal entries of many students, and provides a shared database on the study site for all to use. The collections are an important adjunct to the field guide, for they provide authentic specimens that support the information contained in the field guide. As in a natural history museum, the collection, with its accompanying dates and location information, becomes the permanent and growing data source from which future investigations can be built. Finally, the strategies for observing behavior culminate in the construction of ethograms for recording discrete observable behaviors without attributing motives or explanations. These behavioral spreadsheets provide a permanent record of the observed behaviors of the organisms together with their location, duration, and frequency. Here again, our budding naturalists are contributing to the development of a knowledge base about their study area that can be used to support future investigations.

As the opening vignette suggests, there is no required order in which to pursue these strategies. We have presented them in this book in what we regard as a logical sequence of development, but each strategy is designed to stand alone, could be taught simultaneously with other strategies, or be disregarded altogether in the interest of time. The book is designed to provide a comprehensive program of schoolyard science, but it is also adaptable for shorter term or more casual explorations. In all cases, however, we suggest detailed observation and careful record keeping as the foundation for firsthand learning. We also believe that the whole is greater than the sum of the parts, and that the richest experience will derive from pursuing all of the strategies. From engaging in these observational, data-gathering, and question-generating experiences, students develop a firsthand knowledge of their local habitat that is uniquely their own. Armed with this knowledge they may also become the resident scientists for their school. Less experienced students may bring them their observations and questions, and the evolving, dynamic nature of the local outdoor environment will continue to nurture the growth of this community of learners. Stimulated by this engagement some students may choose to pursue science-related careers. Others may simply come away with a fascination with their physical surroundings and a love of the natural world. Still others may acquire a way of learning that will build confidence in their ability to figure things out for themselves. Most important, through engagement in *Outdoor Inquiries* we hope all students will have an opportunity to develop their abilities *to think and act in ways associated with inquiry.*

Peter Dow

Notes

1. As quoted in E. O. Wilson, *Naturalist* (Washington, DC: Island Press, 1994), 12–13.

2. E. O. Wilson, *Naturalist* (Washington, DC: Island Press, 1994), 166–67.

1

An Overview
How This Book Is Organized

The title of this book, *Outdoor Inquiries*, expresses the dual nature of its purpose. The intent is to provide the teacher with guidance both for how to foster important inquiry skills and how to bring inquiry out into the natural world. Taken separately, each is challenging, but the rich combination of authentic inquiry in a real-world context can result in memorable and rewarding experiences for both teacher and student alike.

Chapter 2

Let's tackle the reservations that teachers have about taking students outdoors first and make the point that the key to success lies in preparation. Chapter 2, Preparing for Outdoor Inquiries, addresses critical first steps to laying the groundwork for conducting a safe and productive field study. Here there is detailed information about how to select a good study site, survey the site, and evaluate the potential natural resources available. An outline of special safety considerations is provided to help the teacher identify hazards and develop safety guidelines with students. Since the proper tools and equipment can greatly enhance a field study, you will also find a full list of suggested materials for carrying out the various projects described in the book.

Chapter 3

After the teacher has done some preliminary scouting at the site, it is time to get students involved in preparations to go outdoors. In Chapter 3, Getting Started with Students, the class is introduced to the exciting idea of doing field studies. They begin by practicing some basic, essential skills indoors. Using objects that the teacher has preselected from the study site, students make close observations and record their first journal entries. They use three simple drawing exercises to help them become more proficient and thus feel more confident about drawing later in the field. Then they complete their journal entries by using descriptive language, measurements, and labels. Students also begin the important practice of writing down questions that arise while they are observing and recording.

When they have completed their first journal entries to their satisfaction, students share and critique their work in small groups. Then, in a whole-class discussion, they develop a list of criteria for good journal entries. As their journal-keeping skills evolve, students will be encouraged to revisit the criteria and add to them.

Since formulating questions is such a necessary skill for inquiry, students will take the time to examine the questions that they recorded in this first exercise. They classify their questions into two categories: questions that they could investigate by making further firsthand observations and questions that would require doing research by using resources such as books or the Internet. This analysis helps to set the stage for later developing viable, investigable questions that students might pursue in the field.

Chapters 4 Through 8

Chapters 4 through 8 focus on five strategies for building a database. Why collect a database of facts and statistics about a place, an organism, or a phenomenon? A solid collection of factual information about your site can be used for any number of purposes—as evidence for supporting an argument, as points for debate, as a way to find patterns and answers, and as a source of ideas and new questions to investigate.

You will notice that we place a high premium on having students collect their own data. In many other learning situations, students use data collected by someone else or simply fill out a ready-made data organizer. They get very little practice going through the rigorous process of figuring out ways to collect, organize, use, and share data, or even thinking about what kinds of data might be important to answer a question. Plus, students have little investment in the impersonal and remote data simply presented to them as fact. Someone else's question and collection of observations about the rain forest do not have the same immediacy as a student's own question and database about the numbers and kinds of spiders that live on the playground. And these observations and data might well give rise to new questions to investigate firsthand, perhaps about the spiders' webs, their prey, and their life cycles.

There are a number of different ways to develop a database and any number of strategies you might use to accumulate information over time. We have chosen five major strategies to guide and help organize an outdoor-based inquiry. These strategies include:

- Using Field Journals
- Mapping the Site

- Creating a Field Guide
- Making a Collection
- Studying Animal Behavior

Each strategy has its advantages, and they are often used in combination to furnish the fullest, most comprehensive view of the study site. The book presents the five strategies in a roughly sequential order and in a roughly increasing order of complexity. So you might use the strategies in the order in which they are presented, from simplest to most complex, or you may choose the one strategy that is most useful for answering a given question.

In each of the chapters, 4 through 8, you will find a model of a doable strategy for collecting meaningful data. Each of these chapters presents:

- background information for the teacher
- a list of suggested materials and equipment
- preparation
- a step-by-step outline of a typical lesson involving that strategy
- extensions to the lesson

At the end of these chapters, there are vignettes that further illuminate the use of each strategy. These include:

- a teacher's story, a voice from the classroom describing how the strategy worked
- a scientist's perspective on why the strategy is important and how it is used at the professional level

Strategy Number 1: Using Field Journals (Chapter 4)

A field journal is the most basic tool of the natural scientist. It engages the journal keeper in making and recording observations, storing evidence, and searching for patterns and relationships. Keeping a journal involves exercising many different skills, including those related to language arts, math, creative arts, and science.

Students keep journals as a pivotal, ongoing activity. Their journals are central to the other four strategies since they are where the primary data are stored as words, charts, maps, illustrations, and measurements. They become the storehouse of questions, new ideas, and thoughts about possibilities for further investigations.

In the sample lesson in Chapter 4, students are presented with an overarching question pertaining to how many different kinds of plants grow in a lawn, and then predict the numbers and types of plants they expect to find in the schoolyard lawn. Outdoors they work collaboratively to survey the actual numbers and types of plants growing in the lawn and record the data in their journals.

As a follow-up, students refine their observations and analyze the class data. They also develop new questions of their own as possible avenues for further field studies. The lesson serves to reinforce journal-keeping skills and gives students practice in collecting, organizing, and interpreting their own data.

Strategy Number 2: Mapping the Site (Chapter 5)

In Chapter 5, students are again presented with an overarching question, this time one that is bigger and more comprehensive: what is out there, and where are these features relative to each other? They are asked to consider the site as a whole and create a map of what is actually there. In the process, they become more aware of the resources in the area and will begin to generate more questions of their own. Students usually enjoy the challenges of mapmaking, and teachers appreciate the opportunities for students to apply skills in math and measurement, scaling, plotting coordinates, and drawing to a real-world situation.

In addition, the map that students produce will become a handy reference tool later when students work with some of the more complex strategies. For example, if students create a field guide to the schoolyard or make a collection, the map will help them locate their specimens more precisely. If they plan to do animal behavior studies, the map is a useful tool for plotting the locations in which animal activities take place.

The sample lesson in this chapter moves students through a logical progression. First they try to create a map of the site from memory, and then, through comparison and discussion, come to realize that they need to go out and see for themselves. Outdoors they work in teams to map portions of the site, and later, back in the classroom, they collaborate to produce one consolidated class map.

We have provided examples of different styles of mapping such as the survey, the grid, and the transect maps so that there is a choice of techniques to consider. The survey map involves the initial assessment of the site, and is the type teachers use most. A grid map is often used for taking more precise measurements and dividing the plot into manageable areas. When using a transect, students lay a straight line along which they take measurements and make observations to obtain a scientific sampling of the site.

Strategy Number 3: Creating a Field Guide (Chapter 6)

To keep interest high and to add a new aspect to students' data collection strategies, have the class create a field guide to the specimens inhabiting their unique study site. In creating a field guide, students will initially have to depend on data previously collected in their journals and on the class site map. They may find that their original records are not as complete or as detailed as they could be, and so may conclude that they need to revisit the specimen itself to collect more data in order to answer a question.

At this stage, students are actively formulating questions, and are more than ready to select one to investigate, one that involves an organism that is of special interest to them. The first step might be to identify the organism and to place it in context by using a commercially available field guide.

Students have used field guides from the first lesson, and should be familiar with how they work. Now students take a closer, more critical look at the commercial field guides to develop a format suitable for the class field guide. Once the class comes to consensus on the criteria for their guide, they return to the specimens outdoors to collect more precise data. Then they create their field guide entries to include important common elements: a line drawing with labels, a written description of the organism, contextual data, and other interesting or distinctive features.

Used to best advantage, the field guide is an ongoing data collection strategy. Students can continue to add new organisms over time, and might even bequeath their guide to the site to next year's class. The project provides a multitude of cross-curricular opportunities in areas such as science, math and measurement, social studies and geography, language arts, drawing, and photography.

Strategy Number 4: Making a Collection (Chapter 7)

In much the same way that a field guide captures a site at a particular moment in time through verbal and pictorial representation, a collection also reproduces the essence of a site, but by using real objects. Objects are the physical evidence that represent the data. As with the field guide, making a collection can become a long-term project.

Many young people take great pleasure in amassing collections of things that interest them, whether natural or otherwise, and so they will be genuinely interested in collecting physical evidence from their study site. In fact, one of the teacher's challenges may be in setting limits on the size and scope of the class collection. The chapter contains guidelines to responsible collecting to help with this common problem.

The sample lesson first engages students in making a plan for collecting. They discuss what to collect and why, as well as how to collect and the tools they will need. Then they apply their decisions to a plant-collecting field trip. Upon returning to the classroom, they identify their specimens, preserve and mount them, and add pertinent data such as where and when collected, distinguishing characteristics or marks, typical range, habitat, and other significant notations.

Strategy Number 5: Studying Animal Behavior (Chapter 8)

Students are naturally drawn to the study of animal behavior, and have no doubt amassed many questions on the topic during the course of their previous excursions outdoors. The first steps in conducting an animal behavior study would therefore be for students to review their own questions and select those that are pertinent.

It's important to recognize the common pitfalls in studying animals, such as the tendency to anthropomorphize and ascribe human emotions or motivations to animals. One way to avoid this temptation is to record the observable behaviors without trying to interpret them. So, for example, a scientist might describe an inactive squirrel as lying still for ten minutes with eyes closed, but would not speculate that it was feeling sick or daydreaming. Several activities help students to break animal activities down into their components and to practice describing them in precise, objective language.

Then students are asked to call upon organizing and planning skills as well as language arts skills to develop their own tool (such as a record-keeping chart or an ethogram) to keep track of one or two behaviors in the animal of their choice. They begin by defining the behaviors they will monitor, and for how long, so that the result will be a kind of time map that shows how often an animal exhibits the behavior during a specified period of time.

Data collection may continue for quite a while, for it is important to observe and record repeated instances so that a pattern begins to emerge. As the observations go on, students may become aware of more, related behaviors that they want to include in their study, and thus expand upon or modify their original plan.

At the end of the study, students analyze their data and attempt to draw conclusions. Each team gives a public presentation to discuss their findings, tell about their plan and their methods for collecting data, the challenges they encountered, and the new questions that arose.

Chapter 9

A good question is the linchpin on which scientific inquiry turns. Chapter 9, Finding Investigable Questions, traces the development of questioning skills through the use of the five strategies. We present the case for leading students through deliberately and carefully choreographed activities in which they first explore and help to define a range of questions, both those that are investigable through firsthand actions and those that may be answered by making use of resources. Once students recognize the differences between the two, they are better able to select and then develop their own investigable questions.

As students continue to observe and record data using the five strategies, more and more questions emerge, and since these questions arose from their own experiences, they are all the more compelling. Students have generated their own need to know, and are more highly motivated to generate the means to find out. Plus, through practice in guided explorations, they have acquired the skills to work more independently to take responsibility for asking and answering their own questions.

Chapter 10

Why teach inquiry-based natural history science? In the closing chapter, we discuss the differences and similarities between natural history science and lab-based experimental sciences, and talk about where to take inquiry next.

At the back of the book, you will find the following appendices:

- Alignment with Standards, a chart that shows how *Outdoor Inquiries* supports the National Science Education Standards (Appendix A)
- Resources and References, including a list of vendors, websites, and reference materials (Appendix B)

2

Preparing for Outdoor Inquiries

Advance planning for a long-term outdoor investigation is a crucial first step. A few days spent on selecting a site, taking a survey of your outdoor resources, making a preliminary map, thinking about safety, and collecting tools and equipment will have a big payoff.

Criteria for a Good Site

Good field study sites need not be exotic or far away. You may find a good study site right outside your classroom door, or you may have to work a little harder and scout the neighborhood for an appropriate location. If your schoolyard is not suitable, consider a nearby park, a vacant lot, a nature center, or a cemetery. Just about any safe, easily accessible site has possibilities, so even if your site doesn't meet all the criteria, it will still provide ample opportunities for investigation. Here are some criteria to keep in mind as you evaluate your site:

- **Safety:** Check the site for potential hazards, such as dangerous debris, holes, poisonous plants, or vehicular traffic. Please see the section called Special Safety Considerations for Working Outdoors for more information.
- **Accessibility:** The site should be easily accessible, within walking distance if possible. Remember to factor in the travel time when you plan your field trips. Anything over ten minutes is probably not practical for frequent field trips. It's also best if you can get there all year long.

    You might also consider a site that students can see from a classroom window. The advantage is that they could set up a feeder for

FIGURE 2–1. This unmaintained area, bordered by a parking lot and a sagging fence, has both sun and shade, weeds, trees, climbing vines, rocks, and plant litter, providing a diversity of habitats.

birds, butterflies, or ants and monitor it throughout the day, perhaps with the help of binoculars.

- **Size:** You will need enough space for the whole class to work comfortably in the area. At the same time, you will want to limit the size of the area so that you can stay in easy contact with your group.

- **Diversity:** A rich site will include a diversity of life forms as well as a range of man-made features. Ideally, the site would have a variety of plant life such as trees, grasses, and shrubs, as well as dead plant material such as leaf litter, fallen limbs, or rotting wood. All of these serve to attract a variety of animal life. Also look for a range of environmental conditions such as shade and sun, wet and dry areas, mowed and unmaintained areas. Figure 2–1 is an example of a site that is close to the school building yet rich in possibilities.

Some Examples of Good Study Sites

Depending on your location, there may be a wide variety of spaces to consider, but it helps to know what you are looking for. Here are some examples, both big and small, of fairly easy-to-find sites:

- **A Transition Zone:** A transition zone is an area where land use changes from one type to another. For example, look for a grove of trees that borders a mowed athletic field, or a grassy hillside that slopes down into a parking lot, or a flower bed surrounded by lawn. Transition zones often lead to questions about the differences in the plant and animal communities associated with each type of land use.

- **Everyday Walkways:** An easily accessible site might be a strip of sidewalk that has a broad median on one side and a building within twenty feet or so on the other side. The street and the building narrow the focus to a well-defined area. If you select a place that students walk through frequently, it is easy to make quick observations over time and notice seasonal changes.

- **Protected Pockets:** A small, ten- to fifteen-foot piece of land between two buildings can yield information on how plant and animal communities are affected by the sheltering effects of structures. For example, the buildings may shield the area from sun, wind, rain, and pedestrian traffic.

- **Cracks and Crevices:** Minicommunities of plants and animals may find homes in some very tiny places. The crack or crevice where the vertical wall of a building meets a horizontal paved surface is a spot where you may find moisture-loving plants and animals that take advantage of the rain that runs down the side of the building and the debris that accumulates there. Cracks in the blacktop or sidewalk may also provide space for hardy weeds or the entrance to a subterranean ant colony. (See Figure 2–2.)

FIGURE 2–2.   A miniature transition zone such as this is easily accessible yet holds a surprising variety of plants and insects thriving in the protected pocket.

- **Tree Wells:** City streets and other high-use areas often have trees planted in tiny islands of soil surrounded by an expanse of concrete. Small enough to take in at a glance, the tree wells provide an opportunity to look at drainage patterns and weeds that inevitably colonize the area beneath the tree.

- **Sky Watching:** You might spot a narrow corridor or courtyard where a small chunk of the sky is clearly visible. Use it to watch bird or insect flight, look for wind patterns, and map temperature changes. Or watch different kinds of clouds and time how long they take to cross the space.

- **Water:** Any kind of water—a puddle, a drippy downspout, or an overflowing drinking fountain—increases the possibility of biodiversity. For example, a low, wet area of lawn is a mystery waiting to be explored. It might lead to questions about why the area is wet, how the moisture affects plant communities, what kinds of creatures are attracted to the water, and the chemical composition of the water itself.

  A storm drain is another potential asset. The pit or well below the drain captures sediment and holds water between rainstorms. Students may discover both plants and animals living in this artificial pond. What kinds of collecting devices would students have to invent to fit between the slits in the grate?

Survey the Site and Make a Preliminary Map

Reserve several pages in your own field journal to take notes and make a preliminary map of the potential site. It will save you considerable time later if you take the time now to develop these useful reference tools.

As you walk the site, jot down information about the resources it holds and the possibilities there for student investigations. You might record some of the following: the numbers of different kinds of trees, bushes, and other plants; animals sightings; evidence that animals use the space, such as nests, tree cavities, or chewed vegetation; differences in rocks and soil; water and how it travels over the site as evidenced by puddles or erosion.

In addition to the resources, you will want to define the space where students will work. Think about whether there are natural boundaries or if you will need to use cones or flags to indicate the perimeters of the site. And don't forget to take note of safety hazards so that you can alert students to the potential dangers before you leave the classroom.

Make a preliminary map of the site for your own use. The emphasis here is on "preliminary." In Chapter 5, Mapping the Site, you and your students will make more detailed and more accurate maps. This one is just a convenient way to jog your memory and to help you in planning.

Another approach would be to combine the mapping with the survey of resources. That is, after you sketch the map, pencil in your notes about resources and where they are located.

When in the planning stages, it is very helpful to know how far it is between areas or how long it takes to travel from one spot to the next, so try to include approximate distances and travel times on your map as well.

A Final Note About Site Selection

Don't be discouraged. You really don't need to go to the rain forest to do a topnotch field study. Even what appears to be a barren, blacktopped play yard can yield a surprising amount of data. Check the cracks for tiny plant life, seeds, and ants. Look around the edges of buildings for plant debris and worms living underneath. Trace around the puddles and time their evaporation. Measure the temperatures. Track the path of shadows. Watch a spider spinning a web on the swing set. Look up, observe the flight patterns of birds. There's a lot out there if you take the time to observe closely.

Tools for Firsthand Learning

The appropriate tools and equipment can enhance outdoor learning, spark questions, and extend inquiry. Here is a list of basic supplies as well as suggestions for other items you might want to add. You might start small with just a few pieces to test. Later you could consider purchasing class sets of some items such as hand lenses, journals, and collecting bags. Assemble the materials in a sturdy container with a handle for easy carrying. A toolbox or fishing tackle box works well.

The Basic Tool Kit
- one field journal per student (either student-made or commercially purchased)
- hand lenses (at least one per group of four students)
- rulers, tape measures, yard sticks or meter sticks
- a ball of string (for marking plots)
- bamboo barbeque skewers (to use with string to mark a plot or to mark a location you want to return to)
- a compass (to determine the orientation of the site)
- a pair of pruning shears (for collecting woody plant specimens)
- a pair of scissors (for collecting softer, nonwoody plant specimens)
- forceps or tweezers (for picking up small objects)
- a trowel (for digging small holes, to take soil samples)
- a mini-flashlight
- collecting jars, unbreakable, such as yogurt or margarine tubs with air holes poked in the top (for transporting specimens such as live organisms)
- collecting bags, such as resealable plastic bags (for transporting plant material, soil samples, rocks)

- thermometers (for measuring temperatures at the site)
- small, soft paintbrushes (for coaxing insects into containers)

Add-on Items
- small nets (for catching insects or aquatic specimens)
- a pH testing kit (for testing water samples)
- eyedroppers (for collecting small samples of liquids)
- plant labels (for marking and labeling plant specimens)
- latex-free gloves (for protection)
- a digital camera (for recording visual data)
- disposable cameras (for recording visual data)
- a multipurpose tool kit (for the teacher or another adult to use in assisting students)

Items to Make Available Indoors
- field guides to plants and animals of your region
- a microscope
- holding containers (for housing captive guests such as insects, snails, or worms)
- plastic trays from the meat market (for examining or displaying specimens)

Please see Resources and References at the end of the book for a list of vendors and information on how to order materials.

Special Safety Considerations for Working Outdoors

Harness the Enthusiasm

The idea of going outdoors to study will no doubt generate a lot of enthusiasm in your class. It is important to nurture that enthusiasm while at the same time encouraging attitudes and behaviors that ensure that the excursion will be both safe and productive. When you take students outdoors, you remove them from the familiar structure of the school building and the behaviors they know are expected of them. You ask them to modify their way of working, interacting, and even moving about.

Spend some time discussing outdoor safety with your class and then lead them to develop their own guidelines for appropriate behavior. One way to do this is to develop a list of safety rules in class first, and then do a quick walkabout outdoors to test out the rules and imagine how well they will work. Afterward, students may want to modify their list. Keep the safety rules posted, add to them as necessary, and use the list to remind students of appropriate and safe behavior before each field trip.

A Sample List of Safety Rules

It is very effective to involve students in establishing a list of safety rules for outdoor work. Here is an example of a list developed by one class:

- Walk.
- Stay in the boundaries.
- Listen for signals from the teacher. [This teacher used a small bell to let the class know when it was time to reconvene at the predetermined meeting place.]
- Don't touch anything and don't take anything (unless we are collecting that day).
- Leave everything the way you found it. Don't litter.
- Use normal indoor voices. Loud noises will scare away wildlife.

Set a Purposeful Tone

Having fun and enjoying the time outdoors will become a pleasant and positive by-product of the experience, but the main objective of your field trips will always be to collect data of some sort. For example, students may be engaged in creating a map, counting or measuring organisms, tabulating animal behaviors, or collecting specimens. From the outset, establish a purposeful tone for each outdoor work session. Make expectations and goals for the fieldwork clear and explicit before you leave the classroom.

Safety Checklist

While it is impossible to foresee all of the situations you may encounter outdoors, there are a few simple measures you can take to promote a safe and successful field trip. For example:

- Begin by reviewing your district's safety guidelines for field trip activities.
- Before leading the group outside, do some preliminary scouting to select an appropriate study site and establish its boundaries (cones and flags work well). Please see the section Criteria for a Good Site for details on how to select a site.
- Check the study site for potential hazards such as vehicular traffic, bee or wasp nests, broken glass, poisonous plants, chemical residue from pesticide application, or holes in the ground.
- Recruit other adults to help manage the outdoor activities. Develop an emergency procedure plan with them.
- With your students, establish the purpose of the field trip and develop a list of safety rules.

- Set up a buddy system.
- Carry a first aid kit and a cell phone. Program in emergency numbers such as the poison control center.
- Caution students against putting anything in their mouths or touching animals (dead or alive) they might encounter.
- Find out if any of your students have allergies and be prepared to respond in case of a reaction.

Ready to Launch

Now that you have done the preliminary work of preparing for an outdoor investigation—selected a site, surveyed the resources, developed a list of safety rules with the class, and assembled tools and equipment—you are ready to introduce your students to Outdoor Inquiries. In the next chapter, you will work with the primary and most fundamental of tools, the field journal, and get students ready to begin observing and recording the natural world in detail.

Something Extra for the Teacher

It is a treat to catch a glimpse of the naturalist at work, and to admire his powers of observation. What follows is an excerpt of Bernd Heinrich's book *The Trees in My Forest*. Notice how one observation can lead to another, and how one question evolves into a series of related questions.

Natural History Investigation: An Excerpt from *The Trees in My Forest* by Bernd Heinrich

Most of the small trees I found with aphid colonies were tended by at least thirty ants. It was impossible for me to keep track of that many individuals all at once. However, I found a more manageable group of only four or five ants on a two-foot-high quaking aspen sapling along the path to the cabin, and on an afternoon in late July 1996, I sat down beside it to do some serious ant-watching. One of the nice things about watching insects is that, unlike birds or mammals, insects seem to be oblivious to observers. They act as if they are in another world. There is little need to worry that their behavior is affected by your presence (unless they are of the kind that suck blood).

At first, I saw the five ants just running around, alternately palpating aphids with their antennae and then stopping to meet face-to-face with each other. When two ants met they palpated each other's heads as they had done the aphids' rear ends. Important transactions were occurring during these tête-à-têtes. Colony identities were being checked. Food was being transferred. Perhaps ants aren't perfect at discriminations. Maybe to an ant there is little difference between another ant's face and an aphid's rear (even though they surely recognize strange ants). Maybe the aphid's secret of survival with ants is that its rear end mimics an ant's front end because ants routinely share food to inform each other about colony nutrition. The aphids have found a simple solution for ingratiating themselves with a potential powerful enemy and making it work to defend themselves if need be. And they do it cheaply, using their own waste.

Within a half hour, two of the five ants with swollen gaster (abdomen) were running down the aspen head first, leaving in a hurry, ferrying honeydew, the aphid's sweet anal secretion, back to their colony. Two new ants had already run up the tree. A small green snout beetle was blundering along the ants' "highway," the thin tree stem. When the ant met the beetle, the beetle tucked in its legs and instantly fell to the ground like a small green stone. I then noticed a leafhopper on one of the poplar's leaves, and it leaped off into space when an ant came near. To my great surprise, I eventually detected a half-inch-long green inchworm (a geometrid caterpillar). It was motionless and aligned neatly along the lower midribs of a leaf, looking like the midrib itself. The ants did not crawl on the leaf undersides. Attached to the underside of another leaf was a frothy white gob of sticky spittlelike material.

Within the "spittle," I knew, would be a "spittlebug," or "froghopper." Adult froghoppers have wings and can jump marvelously far. Like the aphids, they also suck plant juices. Instead of giving up its surplus of plant juice to feed ants, as aphids do, a froghopper larva exudes and then surrounds itself with a protective froth that repels ants and within which it hides. This froth is not spit at all but an anal secretion. The larva turns this liquid into a froth by blowing exhaled air into it and adding a substance like soap that helps make a lasting lather.

After what seemed like an eventful half-hour adventure at the young aspen tree, I was encouraged to do a repeat observation the following morning. This time I'd keep an even closer watch on individual ants. It occurred to me that ants had come up the tree, tanked up on aphid juice, and left, but maybe that was just an illusion. Maybe what really had happened was that the same five ants had stayed there all the time working the aphids, while special "tanker" ants traveled back and forth between the nest after the tappers gave up their honeydew to their colony-mates.

During my second watch at the aspen sapling there were again five ants in attendance. Four of them were slow-moving and deliberate, patiently "milking" the aphids most of the time, but one individual acted noticeably different. This one, whom I shall call "the runner," was continually jogging all over the aspen sapling, occasionally stopping briefly as if marking time. This ant seemed almost oblivious to the aphids. It never stopped once to palpate them. After about ten minutes the runner engaged in a long tête-à-tête with one of the four resident aphid-milkers. Looking close I saw that it was indeed receiving a load of collected honeydew. Five minutes later the same thing happened. Immediately after this tank-up (I could see that it really was a tank-up because the ant's gaster was by now visibly swollen), it ran headfirst down the poplar shoot and hit the ant trail, presumably back to its nest. I was elated that my hunch about the tanker ants was right.

Ten minutes after the tanker had left, one of the other four ants was also becoming swollen with clear liquid honeydew. I could see its dark hard abdominal segments pulled apart to reveal the clear abdominal contents through its transparent membranes. I could look almost through the ant. No "tanker" ant had come in the meantime to relieve the rapidly ballooning ant, but this ant didn't wait. It, too, eventually ran down the poplar and headed back to the colony. The ants were not as inflexibly programmed as I had presumed.

Getting Started with Students

The scene is set: you have selected a good outdoor site for study, surveyed and mapped it for yourself, and gathered tools and materials for students. The next step is to rehearse the players, to provide students with concrete experiences indoors that prepare them to work productively outdoors.

The Field Journal

For all of the projects described in this book, the field journal is a central component. It's important to provide each student with an individual journal in which to record data, plans for research, predictions, new ideas, and new questions.

Field Journal or Science Notebook?

Some science educators make a distinction between *journal* and *notebook*, and may prefer one term or the other. For purposes of consistency, we will use the term *field journal* to convey the idea that this is a personal record of one's work in a long-term study of natural history.

The journal itself may take different forms. There are some excellent journals available commercially, or you may have students create their own folders or loose-leaf binders. We recommend that the journals be:

- Durable. Students will be taking them outside.
- Rigid enough to support writing and drawing. If not, plan to have students use clipboards, pieces of cardboard, or books for support.

- Composed of both blank pages for drawing and lined pages for writing, with a section dedicated to recording questions.

Since the field journal is such a critical element, you will find several other chapters that discuss its use. In this opening activity, students will make their first entries, share them with each other, and begin to develop criteria for good recording. They will also begin to use their journals as thinking tools, places where they record their ideas, voice their thoughts, and ponder new questions. In subsequent activities, they will refine their recording and thinking skills as they continue to practice them.

Using Preselected Objects from the Study Site

In the first activity, students get an enticing glimpse of the field study to come by examining carefully selected objects taken directly from the study site. The objects themselves provide a snapshot of the scene outside. Later, when students encounter similar objects at the site, they will have a sense of familiarity and can then focus on putting the items into a larger context. For example, if a student has drawn, labeled, measured, and verbally described an acorn, it will be satisfying to discover the same kinds of objects in the autumn, lying under an oak tree, surrounded by oak leaves, and perhaps being gnawed on by a squirrel.

What kinds of things might you collect for students to examine? Small inanimate objects work well for beginners. They remain motionless, and students can record them life-sized. Trees and their parts make especially good subjects because they are large, stay in place over time, and reflect seasonal changes. Animal specimens are arguably more exciting, but they are also more challenging. And remember, if you decide to include living creatures, it will be important to provide for their needs while they are in the classroom, and then to return them to their habitat as soon as possible (Figure 3–1).

Here is a sample list of natural things you might collect from the site:

- plants and plant parts, such as grass, leaves, twigs, flowers and buds, seeds, nuts, cones, fruit, bark, thorns, galls, rotting wood
- rocks, coarse dust, fine gravel, soil, sand, mud
- vials of water collected from puddles, a pond, or a downspout
- living specimens, such as arthropods, worms, or snails

You will need one object per student. Multiples of the same objects are fine; they encourage students to compare and contrast.

Drawing Exercises

Drawing can be a stumbling block. It helps to let students know that they are not expected to produce a work of art, but rather to use drawing as a means of recording data about what they actually see. Beginners may struggle at first,

FIGURE 3–1.   Before recording, a student makes close observations of preselected objects from the study site.

but will soon notice that drawing becomes easier with practice, and that the drawings themselves improve dramatically over time, becoming more and more clear, accurate, and detailed.

Contour drawings are common foundational exercises that help improve drawing skills. Contour drawing encourages students to concentrate on close observation. By slowing down and focusing on what is actually in front of them, students short-circuit conventional habits of representation and begin documenting what they see more faithfully. While taking the time to practice these contour exercises, students will also improve their hand-eye coordination. Some teachers have found that an extra ten to fifteen minutes in the beginning devoted to drawing exercises increase student comfort level for recording later in the field.

Three Drawing Exercises

Exercise 1: Blind Contour Drawing (two to five minutes)
Students keep their eyes on the object and draw it in a continuous line without lifting the pencil from the paper.

Exercise 2: Modified Contour Drawing (two minutes)
Students again draw the same object in one continuous line, but this time they are allowed to look periodically at the paper.

Exercise 3: Finished Drawing (five to seven minutes)
Students use one of the sketches to make a more detailed drawing of the object. They add written notes, measurements, color, and labels.

Complete directions for presenting these exercises to students appear in the lesson plan that follows. (See Figure 3–2 for an example of contour drawings.)

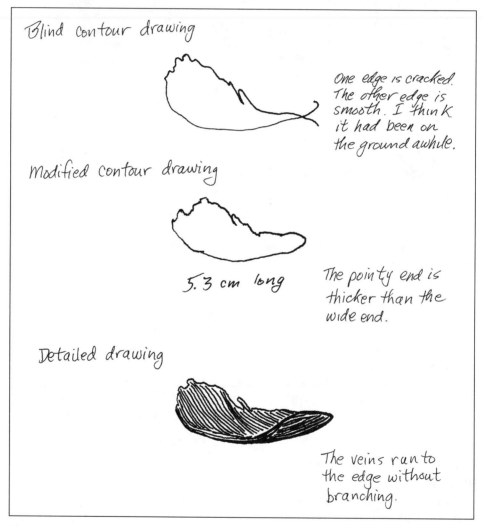

FIGURE 3–2.   Examples of a blind contour, modified contour, and finished drawing of a maplewing

The Activities

The Lesson at a Glance

Students:

- practice drawing using preselected objects
- make their first journal entries
- share and critique their journal entries; establish criteria for good entries
- analyze types of questions
- learn to use a field guide to identify specimens

Materials

- one field journal per student
- colored pencils
- rulers or tape measures
- hand lenses
- an assortment of preselected objects (please see Using Preselected Objects from the Study Site)
- chart paper and markers
- one copy of A Sample Student Journal Entry (Figure 3–3)
- overhead projector
- field guides to your region
- one copy of How to Read a Field Guide (Figure 3–4)
- access to the Internet

Getting Ready

1. From the site, collect objects representative of your local environment, one or more objects per participant. Try to provide a wide range of specimens. Please see the section Using Preselected Objects from the Study Site, from this chapter, for ideas.

2. Choose the kind of field journal your students will use and be ready to provide these. Please see the discussion, The Field Journal, to help you decide.

3. Prepare two charts, one labeled *What to Include in a Field Journal* and the other labeled *Questions*.

4. Make blackline masters of A Sample Student Journal Entry and How to Read a Field Guide (Figures 3–3 and 3–4) to use on the overhead projector.

Warm-Up Activity Indoors: Introduction of Field Journals and Tools

Recommended Time: thirty minutes

1. Launch the Outdoor Inquiries field study. Let students know that they will soon be going outdoors to begin an investigation of a site, and briefly describe some of the projects and activities, such as journal keeping, mapping, creating a field guide, collecting, and studying animal behavior. But first, it will be important to get ready to work safely and productively outdoors.

2. Explain to students that one of their most important tools will be the field journal. Then distribute the journals (if commercially purchased) or the materials for constructing them. When the journals are ready, discuss some of these questions. Record student responses on the chart you have prepared for this purpose.

   * Why keep a field journal?
   * What kinds of information might you record on the lined page? How should we set up that page (if you are creating your own journals)?
   * What kinds of information should we include on the blank pages for drawing?
   * How might you use the information you record?

   Keep the chart posted so that students may review it and make additions later.

What to Include in a Field Journal

At the beginning stages of making journal entries, students should mention some of the following criteria. Later, as they become more experienced, they will be able to add even more.

On the Drawing Page

- a detailed drawing of the object that shows how it really looks
- labels on the drawing to point out features, such as parts or markings
- measurements

On the Writing Page

- a good description of the object that might include size, texture, color, shape, or other interesting observations
- new questions that came up

3. Show students the tools and equipment you have gathered for their outdoor work. These might include hand lenses, rulers, tape measures, and field guides for your region. Ask them how these tools will be useful in collecting data about their outdoor site.

4. Then display the preselected objects you have collected from the site. Lay the objects out on a large table or on the floor and have students gather around. Explain that the objects all come from the area that students will be studying, and that taken together, they give a kind of snapshot of the site.

5. Either let each student select one object or distribute the objects randomly. Also distribute hand lenses and measuring tools, and place colored pencils where students may pick them up later when they are ready to refine their drawings.

6. For several minutes, allow students to enjoy a period of free observation. Encourage them to use hand lenses and to discuss their objects with each other. Then get ready to record.

Three Drawing Exercises

Recommended Time: ten minutes for the three drawings and thirty minutes to complete the journal entry

1. If your class is at all hesitant about drawing, the three drawing activities are a fun way to introduce students to that part of the journal. Give students these instructions:

Exercise 1: Blind Contour Drawing

Look at your object closely. Find a starting point and keep your eyes only on your object. Draw in a continuous line everything you see. Don't peek at your paper. Don't lift your pencil or stop until you have drawn all of the outlines, lines, or markings on your object. Go slowly. Let your eyes "trace" the contour of your object. (Allow about two to five minutes.)

Exercise 2: Modified Contour Drawing

Reposition your object if you like. Draw with one continuous unwinding line, as before, but this time permit yourself to look periodically at the paper. (Allow about two minutes.)

Exercise 3: Finished Drawing

Use one of your sketches to do a more detailed, finished drawing of your object. Add written notes of the object's size, texture, color, or other interesting features. Use colored pencils and measuring tools to refine your entry. (Allow five to seven minutes.)

2. Once students are satisfied with their finished drawings, ask them to complete the journal entry by adding written descriptions, measurements, labels, and their thoughts and questions about the object.

Sharing and Critiquing Journal Entries

1. Arrange students in small groups and ask them to share their journal entries within each group. Ask them to discuss and take note of the following:

   * What do you think you did especially well in your journal entry? Share that with the group.
   * What did other people in the group do well?
   * What are some of the things the group found difficult to do?
   * Did you all include the same kind of information? Could you have added more to your drawing? To your written description?
   * What kinds of questions did people record?

2. Bring the whole class together to report on their discussions. Then return to the chart that students created about elements to include in a journal entry and have students refine their earlier ideas. Ask:

   * What else can we add to the list of what makes a good journal entry?

3. Show students the overhead of A Sample Student Journal Entry (Figure 3–3) and have them critique it. Use some of these prompts:

   * What do you notice about this journal entry?
   * Does it include the elements on our list?
   * Does it include some elements we didn't mention? What can we add to our list?
   * What did the student do especially well? What might the student have done better?

Analyzing Different Kinds of Questions

1. Mention that questions are an important product of science investigations. They help us clarify our thoughts and lead us in new directions for research. Focus attention on the chart labeled *Questions* and invite students to contribute the questions they recorded as they worked. List these on the chart.

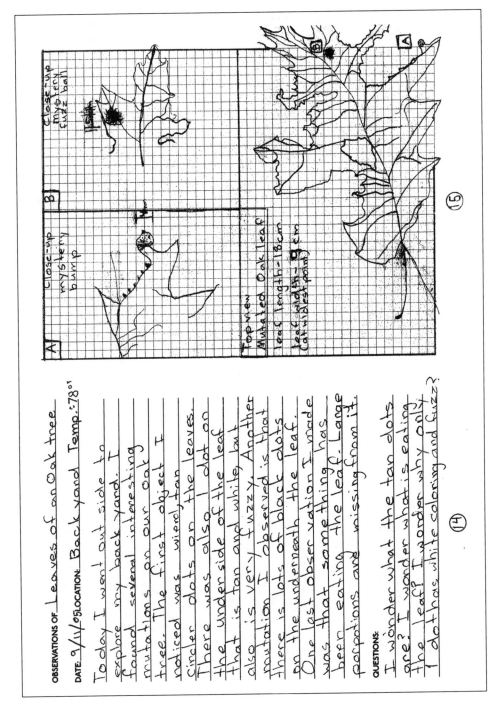

FIGURE 3–3. A sample student journal entry

2. Briefly analyze the questions on the list. Highlight different types of questions; you might identify them with different-colored markers. Not all questions will fit neatly into categories, so there may be some with no notations next to them and some with more than one notation. Ask:

- Which of these questions could we answer ourselves by doing more direct, firsthand observations?
- Which questions could we answer by doing research in books or on the Internet?

Let students know that in upcoming investigations, they will be engaged in trying to determine answers to their own questions.

Examples of Questions

Here is a sampling of different kinds of questions that one class developed about their plant specimens:

Questions We Could Answer by Firsthand Observation

Where could this plant have come from?

What might this plant look like during the winter?

Does this plant grow in sun or in shade?

What eats this plant?

Questions We Could Answer by Doing Research

Why are there hairs on this plant?

Why are the roots twisted?

Does this plant grow in any other part of the world?

Is it poisonous?

Notice that many questions that require research begin by asking *why* something is so. These are often difficult (and sometimes not safe) for students to answer by observation alone. For more information on investigable questions, please see Chapter 9.

Follow-Up: Identifying Specimens

1. Provide students with field guides and/or access to the Internet so that they can work on identifying the preselected objects they have just examined.

2. Since students will be working to identify and classify organisms in future projects such as mapping, making their own field guide, and

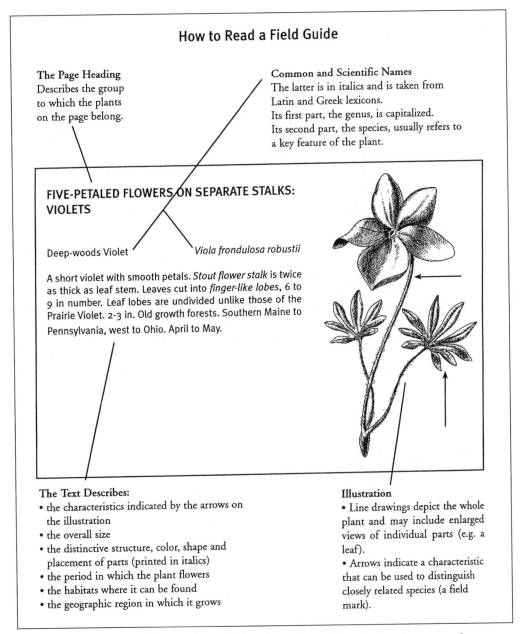

How to Read a Field Guide

The Page Heading
Describes the group
to which the plants
on the page belong.

Common and Scientific Names
The latter is in italics and is taken from
Latin and Greek lexicons.
Its first part, the genus, is capitalized.
Its second part, the species, usually refers to
a key feature of the plant.

**FIVE-PETALED FLOWERS ON SEPARATE STALKS:
VIOLETS**

Deep-woods Violet   *Viola frondulosa robustii*

A short violet with smooth petals. *Stout flower stalk* is twice
as thick as leaf stem. Leaves cut into *finger-like lobes*, 6 to
9 in number. Leaf lobes are undivided unlike those of the
Prairie Violet. 2-3 in. Old growth forests. Southern Maine to
Pennsylvania, west to Ohio. April to May.

The Text Describes:
• the characteristics indicated by the arrows on
  the illustration
• the overall size
• the distinctive structure, color, shape and
  placement of parts (printed in italics)
• the period in which the plant flowers
• the habitats where it can be found
• the geographic region in which it grows

Illustration
• Line drawings depict the whole
plant and may include enlarged
views of individual parts (e.g. a
leaf).
• Arrows indicate a characteristic
that can be used to distinguish
closely related species (a field
mark).

FIGURE 3–4. Field guides provide condensed information in words, pictures, and
measurements to help in identification.

collecting, it is useful to introduce them now to how a field guide works. Show the overhead called How to Read a Field Guide (Figure 3–4) and lead students through a discussion of the types of information they will find there. In this example, they will examine a drawing and a written description of a violet. Notice:

- the page heading that describes how the plant is classified
- the common and scientific names of the plant and how the names are written
- the text that gives information about overall size, distinguishing marks or structures, time of flowering, habitat, and region
- the illustration with arrows that point to distinguishing characteristics of the plant

3. Although it is not the way most naturalists would work—they would take the guides out into the field with them—ask students to begin the identification process by using only the data they have recorded in their journals. They will need to depend solely on their own observations as they recorded them. The process will help make students more aware of the details they may have missed.

    Then let students look back at the real objects, if necessary, to complete the identification.

4. When students have identified their specimens—or have come to a temporary dead end—have them add that new information to their journal entries.

Follow-Up: Reinforcement

In a closing discussion, take a few moments to review what students have learned about the importance of recording data clearly, accurately, and in detail. They may have come to a new understanding of how their observations will help them to identify specimens later. As a result, they may also want to add more to the class-generated list of what to include in a journal entry. Use some of these topics to guide the discussion:

- Were you able to identify your object using only your own journal notes? Why or why not?
- What was missing? What else could you have recorded?
- What part of the recording do you think worked best for you in helping you to identify your object? Was it the writing, the drawing, or the measurements? Or was it important to have done the combination?
- What do you think you will try to do better the next time you record data?

4

Using Field Journals

Rehearsals are over, and it's almost time to head outdoors. Students have practiced drawing, made their first journal recordings, critiqued them, and created a list of criteria describing good entries. Now they are ready to take observation and recording to the next level by practicing their new skills outdoors and in context to answer a real question.

But before students leave the safety of the classroom to apply their newly acquired skills, it helps to review the list of criteria they established, both to remind students of what they might include in their journals and to expand upon the list, if appropriate. Be sure to take the time to review the list. It helps to bolster student confidence when they know what is expected and how to proceed.

Most would agree that activities using journals should include the following elements, although not all of them form part of the written record:

- *A clearly stated question to be investigated.* For this first excursion, the question is provided.

- *A prediction, with reasons given.* Predictions may evolve through class discussion of the question and may become part of the class record of charts or may be recorded in individual journals.

- *A basic plan.* Depending on the complexity of the question, the plan may be informal and involve only discussion or it may be a written outline of what students will do, what equipment they will use, and how they will record data.

- *Records of observations in words, labeled pictures, and numbers.* These are the heart of the record keeping and the important first steps toward building a database.

- *Conclusions based on data.* These may be written or derived through a closing discussion.

FIGURE 4–1. A student records data in his field journal outdoors.

- *New questions.* It is usually necessary to prompt students often during an investigation to keep an ongoing log of questions as they arise, until the practice becomes habitual.

One more thing: safety. If you haven't already done it, please check the sections on safety in Chapter 2 and take some time to make a list of safety rules with your class. Post the list of rules so that you can remind students of appropriate behavior just before you go out. As students find the need for more rules—and they usually do—have them add to the list.

The investigation that follows uses the plants in a lawn as a focus. At first glance, a lawn appears to be made up of just one type of plant, grass. On closer observation, however, we discover a surprising diversity of life, both plant and animal. This diversity is closely linked to environmental conditions as well as to human activity in the area. For example, a sunny lawn area used for play will support different organisms than a shady, low-traffic border area at the edge of woods.

A lawn is a human-controlled environment, and without maintenance it follows a process of succession until a forest (or desert) community finally replaces it. Plants in the lawn not only compete with each other for limited resources, but must also survive human activities. In the following lesson, students examine plants that have adaptations that enable them to be successful despite mowing, foot traffic, and pesticide application.

The Investigation: What's in a Lawn?

The Lesson at a Glance

Students:

- discuss and then make predictions about the variety of plants they will find in the lawn
- plan how to conduct the investigation
- record data in the field
- share their findings
- analyze the class data to come to conclusions
- report on new questions

Materials

- one field journal per student
- hand lenses
- chart paper and markers
- measuring tools, such as tape measures, rulers, yardsticks, or meter sticks
- field guides to plants in your region
- colored pencils
- plot markers for each team, such a lengths of twine, rope, unbent clothes hangers, or hula hoops

Getting Ready

1. Prepare a chart for the opening discussion. Label it *How Many Different Kinds of Plants Grow in a Lawn?*
2. Prepare another chart to use after the field trip and label it *Findings*.
3. Gather materials listed above.
4. Divide the class into study teams of three or four students each.

Indoors

Recommended Time: one class session

1. Discuss the different ways the class has prepared for an outdoor investigation. Ask:

   - Do you feel ready to do a good outdoor investigation? Why or why not? What have we done so far to get ready? What else do you think we could do?
   - How can we make sure that we work safely? What rules should we keep in mind?

2. Then mention that as the study goes on, students will be developing their own questions to investigate. But for this first field trip, they will all be gathering data to try to answer the same big question. Focus attention on the chart called *How Many Different Kinds of Plants Grow in a Lawn?* and ask students to share their ideas. Record students' ideas on the chart. Use some of these prompts to guide the discussion:

   • Make a prediction. How many different kinds of plants do you think we will find growing in the lawn? Why do you think that?

   • Do you think there will be more of one kind of plant than another? What plant might that be? Why do you think so?

   Leave the chart posted so that students can compare their findings with their predictions after the field trip.

3. Make a plan. Discuss the best ways to record data to use as evidence in answering the big question. Let students know that they will be working in teams, and that each team will select its own plot to survey. Students might suggest some of the following steps:

   • Pick a plot and mark it out. (See suggestions for plot markers in the Materials section.)

   • On the blank page of the journal, make a labeled drawing of each new plant found in the plot. On the lined page, describe the plant in words and numbers.

   • Tally the number of each kind of plant found in the study plot.

   • Name the plants, even if you don't know what their common or scientific names are, so you can talk about them together later.

Planning

Ideally, students would take the lead in making plans for the activity, and come up with as many of the planning steps as possible themselves. At this early stage they may need some prompting and some help in thinking through the steps.

4. Have students record the question in their journals. Then form the work teams, distribute materials, do one last check of the safety rules, and head outdoors.

FIGURE 4–2.   The student has used a hula hoop to mark the area of the
lawn she will study.

Outdoors

Recommended Time: one class session

1. Take students to the study site and point out the boundaries of the
   area where they will work. Together, walk the site, and ask:

   - What do you observe?
   - Do all areas of the lawn look the same to you?
   - Are you noticing different plants in the lawn?
   - What area of the lawn would you and your team like to study? Why?
     Allow teams some time to select their study plot and mark it.

2. Then set students to work. Remind them to note the basic information in their journals such as time of day, date, and weather conditions before they begin collecting data to answer the question about
   how many different kinds of plants grow in the lawn.

3. As students work, circulate among the groups and use some of
   these focus questions.

To get students started:

- How have you divided up the work? Does everyone on the team have something to do?
- What do you notice about the plants in your plot? What are the similarities and differences?

To prompt students to look for more detail:

- What have you observed about the size, shape, color, or vein patterns of these plants? Are there any flowers, fruits, or seeds?
- What have you measured?
- What have you counted?

To encourage record keeping:

- What is the question we are all trying to answer?
- What data have you recorded so far as evidence to help answer the question?
- Did any new questions come up? Where did you write them down?

4. When students have finished recording, reconvene the group, check that the study site is as you found it, and return indoors.

Indoors

Recommended Time: one or two class sessions

1. Give students some time to refine their journal entries. They may want to complete a drawing or add information to their written descriptions. Provide colored pencils for students to enhance their illustrations. Make available some field guides to the plants of your region so that students can identify their specimens.

2. Hang up the chart labeled *Findings* and ask students to spend a few minutes discussing their data with their teammates. Mention that although each team has surveyed only a small portion of the lawn, together they will put together the pieces of the jigsaw puzzle to get a bigger picture of the whole area.

3. Then ask each team to report their data on how many different kinds of plants they found in their study plots, and how many of each kind they counted. Record each team's data on the *Findings* chart.

4. Analyze the data. Challenge students to use the combined class data to answer the original question: How many different kinds of plants grow in the lawn? Data analysis is a difficult task for many students, so it is good practice to begin with a fairly simple,

straightforward set of numbers such as this, and one that they understand because it is their own.

5. There was also a subquestion posed: Do you think there will be more of one kind of plant than another? Look at the data with the class again, and try to figure out together if one type of plant predominated.

6. Go back to the original chart on which students recorded their predictions. Have them compare their predictions with their findings, and draw some conclusions. Ask:

   • What is a prediction? Can a prediction be wrong? Make the point that a prediction is a record of your thinking at a particular time, so it is neither right nor wrong, but you can collect evidence to prove or disprove a prediction.

   • How did your predictions match up with your findings?

   • Did the evidence support your prediction? Why or why not?

7. Ask students to share some of the new questions they recorded as they worked outside. Stress the value of these questions as new avenues for further study. Ask:

   • What new questions do you have now about the study site?

   • How could we answer these questions? Which ones could we answer by further work at the site? Which ones could we answer by doing other kinds of research, such as in books or online?

   Plan to allow students to do some follow-up studies to find answers to their own questions.

Examples of New Questions

As students work on the broad question of how many different kinds of plants live in the schoolyard lawn, many more specific questions begin to arise. Here are some typical examples:

• The plants in our plot are different from the ones right next to ours. What might explain the differences?

• Are there more different kinds of plants in the sun or in the shade?

• What would happen if our plot didn't get mowed for a week? For a month?

• Do any animals use the lawn? What for? Which plants do they use?

• What are the names of all these plants?

• Do these plants live anywhere else in the world?

When students analyzed these questions, they decided that they could probably find answers to all but the last two questions by doing more firsthand observation at the site itself. The questions about naming and about worldwide distributions of plants would best be researched using books or other such resources.

Follow-Up: The Ongoing Use of Field Journals

The field journals will continue to be a central feature of any outdoor inquiry project you decide to undertake next. The lesson on plant diversity in the lawn provided students with the opportunity to use their journals to gather data and to provide evidence to support an overarching question. They then analyzed their data and drew conclusions based on their recorded evidence. Students will continue to develop these skills each time they use their journals.

Students need multiple opportunities to practice and reinforce the new skills. Their interest has probably been sparked by this first investigation, and you can build on that to encourage more record keeping. You may want to offer students opportunities for less structured observational recording in which they develop their own questions and build a database of evidence to support their findings. Given enough experiences with recording, students will come to discover the power of their own storehouse of data.

Something Extra for the Teacher

Below you will find articles written by two experienced journal users. In the first, teacher Gail Grigg speaks from years of experience in the classroom, and discusses the power of using journals with students of all abilities. In the second, Mark Baldwin, a naturalist and teacher at the Roger Tory Peterson Institute, shares his reflections and several of his journal pages to chronicle his evolution as a scientist and the role that journal keeping has played in his career.

Using Journals in the Classroom

Gail Grigg
Teacher, Royalton-Hartland Central School District
Middlepoint, New York

Over the last ten years journals have been the central tool, a constant, in my classroom. Journal keeping promotes a cyclical process of thinking, testing, and communicating that I have come to see as incredibly valuable. Questions lead to insights, which in turn lead to new questions, then new insights, and still more questions. Journals allow my students the freedom to record ideas without criticism and as such they become individual road maps depicting each child's thoughts about a topic. With journals you can *see* what a child is thinking and use those insights to foster more effective communication and offer more focused feedback.

I have seen children's higher-order thinking skills develop over time as their ability to articulate ideas in words and pictures grows. The evolution is idiosyncratic, but the journals offer me evidence of each child's individual progress. When Genevieve began her discussion of "What is a mammal?" she produced an abundance of facts so detailed that no one would doubt her superior intellect. However, on close examination, she was missing how separate facts and concepts were interconnected. While she could offer encyclopedic information, she was unable to tell you why a certain skull had eye sockets on the side of its head and what that might have meant for the animal's survival. By the end of her journaling experience with mammals, the entry had taken on a completely different perspective. When asked, "What is a mammal?" her explanation included the details of classification but continued to explain why those details were so important to the survival of an animal in its particular environment. The evolution of her thinking could be seen in the entries she made between the first and the last. As her mentor, I had asked her questions about what she knew. In turn, she began to question herself and her observations, which led her to make more connections and develop concepts.

Other students were less focused on recitation of facts, but were also much less productive in their initial entries. Joseph's first journal entry began with a simple list of items: "hair, warm-blooded, eats other animals, breathes." The drawing that accompanied his explanation was of a cat eating a mouse. There were no labels and no other details in the picture. As he progressed through the journaling experience, his drawings changed the most. By his last entry, he had depicted the animal that matched his mystery skull (a beaver) in an environmental context. Joseph drew a lodge on a pond showing the underground entrance, with two kits in the lodge and small branches set to the side. He included detailed measurements of the lodge, the pond, and the beaver's head, feet, tail, and body. His writing consisted of simple sentences explaining the picture, but the concept of the relationship between the animal and its environment was very obvious through his picture alone. His progress in the development of "interrelatedness" was just as substantial as Genevieve's.

Much of this deeper, more connected understanding came about by encouraging peer interaction in an environment set up for sharing. In my classroom children are moving, talking, showing, and comparing. They debate freely. The journals become a catalyst for change of thought. They are not neat pages filled with perfect handwriting. They have notes squiggled in the margins, little pictures embedded in larger ones and lots of smudges and arrows redirecting thought. Some pages are wrinkled and some are just plain muddy because the journals were needed outdoors. In fact it was outdoors that students most often asked, "Can I use another page?"

Another thing I have learned through the use of classroom journals is that assessment of science knowledge and growth becomes more relevant to students if they are actively involved in "doing science" themselves. If students are engaged in processing their own meanings, a student's journal becomes a record of scientific growth within the scientific community of the classroom. Julia was certain that all insects had six legs and only six legs. However, her belief was challenged when she collected an ant with only five legs that had been actively participating in its colony. It looked like an ant. It acted like an ant. It lived in the same place as an ant, but how could it be an ant without six legs? Through her journaling, she wrote theories, researched, answered her own questions, put it to the test of the community of learners and arrived at a conclusion, which she admitted she could change *if* someone could prove differently. Her explanation: the ant was handicapped just like some people and it still functioned in its society, just like some people. Using a rubric to assess her thinking, she scored high on all levels.

I am continuing to develop my skills and techniques for assessing journals. I have not found the perfect rubric or the perfect means to assign a grade. I still struggle with giving grades at all. In assessing students' work, I look for evidence of change, advancements in concepts and reasoning. I look for growth in the observation of details and the recording of those observations. I look for vocabulary changes. Finally, I look for questions. There needs to be more questions asked and a change in the kind of questions asked. The questions need to reflect a higher-order thinking skill instead of rote knowledge. I call this the difference between "skinny" questions and "fat" questions.

When my students first began the journaling experience, I used to become very disheartened by their lack of questions. They did not wonder. They were not compelled to explore. They were satisfied with being given an answer, and they seldom questioned those answers. To help them regain that sense of questioning, which is essential to the inquiry process, I model what I want them to do. If asked a question, I answer with one. As I meander around the room, I stop and read journals and ask questions. Not the skinny ones, but fat, juicy ones that may or may not have an answer they can reach. "What would happen if the tracks went into the water instead of out?"

In a world of immediate feedback it is no wonder that some of my best students were frustrated by my lack of answers. But I knew I was on the right track one day while I watched Ryan working within his group that included two children who were educationally challenged. When they asked a question, expecting to be told what to do, Ryan answered, "Well, what do *you* think you should do?"

On the Value of Keeping a Nature Journal

Mark Baldwin
Director of Education, Roger Tory Peterson Institute
Jamestown, New York

Ever since I was a kid I have been a collector of natural objects. I used to fill dusty cabinets with rocks and fossils, mussel shells, and pinned insects. In time

I did away with actual collecting and "bagged" my quarry on lists and check-marks in dog-eared copies of Peterson field guides to wildflowers, ferns, and birds. But I didn't become a naturalist until I started keeping a nature journal.

It happened when, as a graduate student, I signed up for a weekend work-shop in natural history illustration taught by Clare Walker Leslie. Her book, *The Art of Field Sketching*, was to be our text. Considering my lack of drawing skill I bought the book a few weeks in advance and started practicing. The exercises in that book had a remarkable effect. They were visual, methodical, and easy to grasp. Soon, faster than I expected, I was drawing fairly satisfac-tory representations of what I was seeing. And—this was the crux of the whole experience—I found my observation skills improving. Leslie's book continued to resonate long after the workshop. I searched for more books on drawing, nature drawing, and the published journals of natural scientists. I was hooked. I dug into the granddaddy of drawing instruction manuals, *The Natural Way to Draw* by Kimon Nicolaides (who was, by the way, Roger Tory Peterson's draw-ing instructor), and got hooked even deeper. I've been keeping a nature jour-nal ever since.

When I enter the natural world I now "bag" observations. For example, while seeing an Olive-sided Flycatcher for the first time, on May 22, 2002, I had my journal with me to record what I saw. I can now look at my record for that date and see that what I was observing looked like *this*, just a thumb-nail sketch and a few words, the thing as it appeared to me at that moment. (See Figure 1.)

The act of putting pencil to paper in this way does two things very well. First, it directs my attention away from my self and engages me with what is there in front of me. Because the event is caught on the page *while it is hap-pening*, and while the sensory channels are wide open—no symbols, clichés, or preconceptions allowed—this leaves me with a record I can rely on. Sec-ond, the event is caught in my mind in a way that makes its recollection, even years later, vivid and detailed.

Questions propel scientific inquiry. My journal is a place where I can ask questions. They range from the simple—"What is this?"—to the complex—"How has this woodlot changed in the last fifty years, and why?" Sometimes I need to know the answer right away. Other times a question needs further investigation or refinement before productive inquiry comes out of it. Part of the fun of inquiry is serendipity. For example, on the afternoon of April 19, 1992, I was raking the leaves off my backyard strawberry patch when I saw something unusual. It was a tan-colored worm, very slimy, flat and ribbon-like, with a peculiar shovel-shaped head. I thought it was some sort of flat-worm, and went to fetch my journal to record a good close look. This was before the Internet, and my search through my own books and those at the library turned up nothing. Then, a year or so later, I happened upon an ar-ticle describing an animal just like what I had recorded in my journal. It was a shovel-headed garden worm, *Bipalium kewense*. The worm, I discovered, is

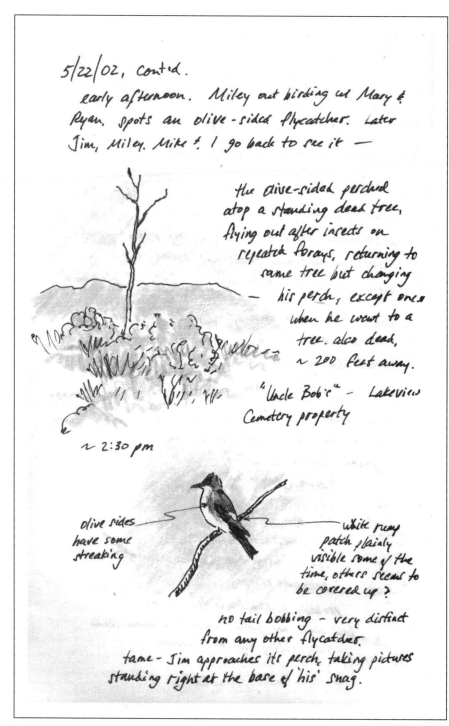

5/22/02, cont'd.

early afternoon. Miley out birding w/ Mary &
Ryan. spots an olive-sided flycatcher. Later
Jim, Miley. Mike & I go back to see it —

the olive-sided perched
atop a standing dead tree,
flying out after insects on
repeated forays, returning to
same tree but changing
— his perch, except once
when he went to a
tree. also dead,
~ 200 feet away.

"Uncle Bob's" - Lakeview
Cemetery property

~ 2:30 pm

olive sides
have some
streaking

white rump
patch plainly
visible some of the
time, others seems to
be covered up?

no tail bobbing - very distinct
from any other flycatcher.
tame - Jim approaches its perch, taking pictures
standing right at the base of 'his' snag.

FIGURE 1. Flycatcher journal entry

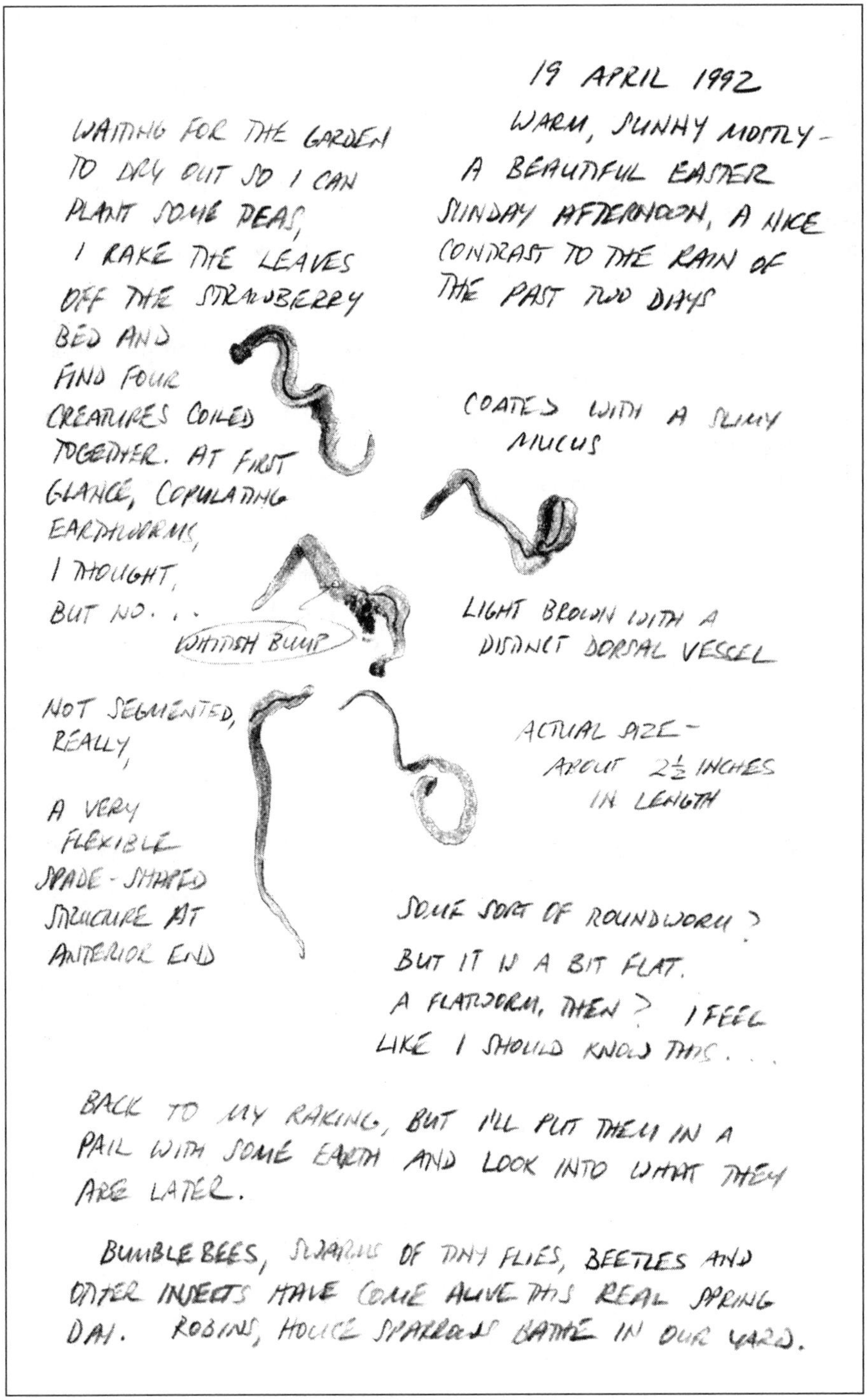

19 APRIL 1992
WARM, SUNNY MOSTLY —
A BEAUTIFUL EASTER
SUNDAY AFTERNOON, A NICE
CONTRAST TO THE RAIN OF
THE PAST TWO DAYS

WAITING FOR THE GARDEN
TO DRY OUT SO I CAN
PLANT SOME PEAS,
I RAKE THE LEAVES
OFF THE STRAWBERRY
BED AND
FIND FOUR
CREATURES COILED
TOGETHER. AT FIRST
GLANCE, COPULATING
EARTHWORMS,
I THOUGHT,
BUT NO . . .
WHITISH BUMP

COATED WITH A SLIMY
MUCUS

LIGHT BROWN WITH A
DISTINCT DORSAL VESSEL

NOT SEGMENTED,
REALLY,

A VERY
FLEXIBLE
SPADE-SHAPED
STRUCTURE AT
ANTERIOR END

ACTUAL SIZE —
ABOUT 2½ INCHES
IN LENGTH

SOME SORT OF ROUNDWORM?
BUT IT IS A BIT FLAT.
A FLATWORM, THEN? I FEEL
LIKE I SHOULD KNOW THIS . . .

BACK TO MY RAKING, BUT I'LL PUT THEM IN A
PAIL WITH SOME EARTH AND LOOK INTO WHAT THEY
ARE LATER.

BUMBLEBEES, SWARMS OF TINY FLIES, BEETLES AND
OTHER INSECTS HAVE COME ALIVE THIS REAL SPRING
DAY. ROBINS, HOUSE SPARROWS BATHE IN OUR YARD.

FIGURE 2. Flatworm journal entry

indigenous to southeast Asia, first became known to science in the hothouses of Kew Gardens in England, and now turns up in North American gardens. They probably came to my garden riding on the strawberry plants I had purchased the previous year from a mail-order nursery. Interesting to note as well, the slimy animals are carnivorous, favoring a diet of earthworms. So now I know *B. kewense* as a pest, and promptly dispatch those I find in my garden, which I still do, from time to time. This knowledge would never have come to me had I not spent the few minutes I did carefully observing and recording in my journal. (See Figure 2.)

I liken my nature journal to a savings account. Recorded observations are similar to deposits; the more I deposit the more knowledge grows, like accruing interest. Plus, I can draw on my simple jottings and sketches and use them to make essays or watercolor paintings. However, unlike withdrawals from a bank account, borrowing from my fund balance in this way can never decrease, but only increase its value.

5

Mapping the Site

Why Create a Map of the Site?

As part of the preparation, you have already drawn up a preliminary map of the site, no matter how rudimentary. In the process, you also became more aware of the educational potential at the site. The map was meant to be an aid to planning and facilitating the first outdoor learning explorations for the class, and a helpful organizing tool as well as a reference document.

For students, mapmaking is an active, engaging project with many cross-curricular links. They will have an opportunity to make real-world applications of skills involving math and measurement, scaling, plotting coordinates, and drawing. Often the process launches students into a productive question-finding mode that leads to further focused investigations.

The map is another type of database, and can later serve as a reference tool when students work on more complex projects. For instance, if students develop their own field guides to the schoolyard (see Chapter 6) the map will function as an important reference tool for locating specimens. If students make collections (see Chapter 7) they will be able to label their specimens more precisely by using the map. And if students conduct animal behavior studies (see Chapter 8) a map will be very useful to plot the locations in which particular animal activities occur.

Data to Include on a Map

Depending on the complexity of your site, you may be able to collect data on a great variety of features and conditions. These might include:

- man-made structures, such as buildings, fences, paved areas, and playground equipment

- soil conditions as evidenced by color, texture, compaction, drainage, and erosion
- ground covers, such as grasses, weeds, vines, or flowering plants
- woody plants, such as shrubs and trees
- changing conditions, such as temperature, precipitation, sun and shade, and wind direction

What Kind of Map Should We Make?

There are many different mapping styles and techniques to consider. Most of the differences have to do with the scope of the map, the level of detail, and the precision of measurement involved in making the map. Let's consider just three types here: the survey map, the grid map, and the transect line. A survey map helps you get a general picture of the whole site, and encourages students to take stock of the resources. Since it is a rather informal tallying and recording, measurements and numbers may not be exact. Both grid maps and transect lines look at smaller samplings rather than at the whole site; both require more precision in measuring and counting.

- **The Survey:** One way to begin is with a rather informal survey of the site such as the one pictured in Figure 5–1.

    Using a simple outline map of the building and the grounds (see the Materials and Getting Ready sections for how to make one), have students take an inventory of the living and man-made features in the area, and record them on the outline map. At this point, you may not want to make precise measurements, but rather get an idea of the big picture. What's out there? What are the possibilities for further study? What kinds of questions arise about the site? The informal survey may then lead to the need for a more complete, accurate, and detailed map.

- **The Grid Map:** After taking inventory of the site, students may formulate questions about some of the features they observed or develop an interest in particular phenomena. This might lead you to select one or two biotically rich or especially intriguing areas of the site for study. In this case you could divide the areas into zones and deploy student teams to cover each zone. Here is where a grid map using coordinates becomes practical as a more precise, mathematical way of defining locations.

    Use stakes and twine to lay out the grid on the site. The size of the site determines how large the grid will be, how many squares it will contain, and how large each square will be. In Figure 5–2, the grid has twenty-four squares; the overall size is sixty feet by forty feet, and each square measures ten feet by ten feet.

    Assign coordinates to each axis to mark off the grid. Alpha-numeric systems work well, where one axis is labeled with numbers and the other with letters—both in ascending order. Each block on

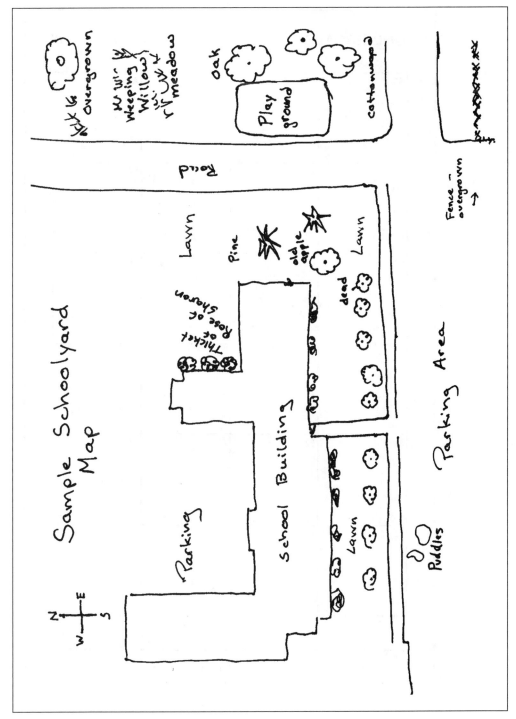

FIGURE 5–1. A sample schoolyard map shows both man-made and natural features.

Gridding the Site   SCALE: 1 QUADRANT =10'

A1	A2	A3	A4
B1	B2	B3	B4
C1	C2	C3	C4
D1	D2	D3	D4
E1	E2	E3	E4
F1	F2	F3	F4

FIGURE 5–2.   This type of grid may be laid out on a site to divide the area into zones and to define the zones to be studied.

FIGURE 5–3.   Setting up a transect line

the grid, A1 for example, will represent a particular location of a particular, predetermined size.

- **Transect Lines:** You can also collect good data by doing a set of line samples called transects. These are sampling areas laid out along a continuous straight line. First you define one axis by stretching a string in a straight line. Then place stakes at set intervals along that axis. From each stake, make a side trip at a right angle from that axis to a set distance such as a hundred feet or a hundred meters. Choose an appropriate sampling distance, such as every foot or every meter. (See Figure 5–3.)

The following Investigation details one possible approach to creating a survey map of your schoolyard.

The Investigation: Mapping the Site

The Lesson at a Glance

Students:

- draw a map of the school and grounds from memory
- work in collaborative teams to map sections of the study site
- consolidate their maps to create one large class map
- share new questions

Materials

- one field journal per student
- measuring tools, such as tape measures, meter sticks, and trundle wheels
- a compass (to confirm orientation on the map)
- a map of the school buildings (see Getting Ready)
- a large sheet of chart paper and markers (for the whole-class map)
- a large sheet of chart paper labeled *Questions*

Getting Ready

1. Obtain a map of the school buildings. Most schools have emergency evacuation plans posted near exits in the classrooms. These show the shape of the buildings and often include some outdoor structures as well. The map is usually a simple outline that is easy to trace. From the tracing, make student copies of the map.

   If you have Internet access, you can obtain an aerial view and a topographical map of your school and its surrounding neighborhood from the U.S. Geological Survey website or through Stanford's geographic information systems (GIS) for individual states.

2. Gather the materials listed above.

Premapping Activities

Recommended Time: one class session

1. Before taking the class outside to map the site, ask students to draw a map of the school and grounds from memory in their journals. In small groups, have them compare their maps with each other.

2. Then distribute copies of the site maps that you have prepared. These usually show only man-made features, and will help give students a sense of the scale of the structured spaces. Ask:

   - What is missing from this map?
   - How could we make a more complete map that also shows the natural features of the site?
   - What kind of data would we need to collect?
   - What tools would we need?

3. Divide the class into teams, and use the site maps to make a plan for which team will map each area. Briefly discuss mapping conventions such as using symbols, creating a legend or key, indicating compass directions, and showing measurements.

4. Arrange to take the class outdoors for several mapping sessions. With the class, review safety guidelines for working outdoors.

Mapping the Site

Recommended Time: one to three class sessions

1. Before going outdoors, remind students of the purpose of the fieldwork: to make a map of the physical and natural features on the
   site. Distribute the copies of the site map that students discussed in
   the previous class session. Ask them to record all their new data on
   the site maps. Review safety rules and gather equipment.

2. Take the class outdoors. Define the boundaries of the study site. Remind
   the class to work with their teams to create as complete and as accurate
   a map as they can within the time they have to work (Figure 5–4).

3. Use the compass to determine the orientation of the site, and have
   groups record a compass rose on their maps.

How to Use a Compass

Notice that your compass needle points in two opposite directions.
In most cases, the needle has one end painted red. The red end points
to the magnetic north. Place the compass in the palm of your hand
or on a flat surface. Wait until the arrow stops jiggling and comes to
rest. Gently twist the compass until the red end of the arrow lines up
with the letter *N* and there you have it—you have found north.

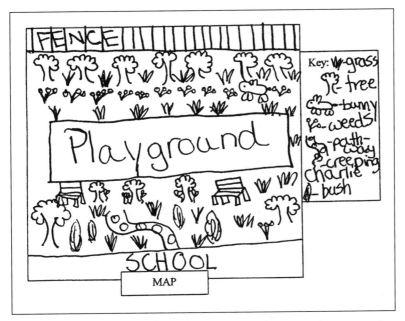

FIGURE 5–4.   This student map shows both man-made features such
as the school, the playground, a fence, benches, and a pathway, as well
as natural features such as several kinds of plants and an animal.

4. As students set to work, circulate among the groups and interject some focus questions. To get students started, ask:

   - How has your team divided up the tasks? What is each person doing?
   - What are you measuring? What tools are you using to measure?
   - What symbols are you using to represent the different objects on the site?
   - Have you created a key or legend for the symbols?

5. Encourage students to include both natural and man-made features. For example:

   - physical features, such as streets, driveways, play areas, fences, flagpole, bike rack, and parking lot
   - natural features, such as trees, shrubs, vines, ground cover, lawn area, flowering plants, boulders, and water

6. When students have made some progress in recording the features of the site, prompt them to make more observations that might point to significant signs of activity. At this time, questions often arise that could lead to further investigations. Try these prompts:

   - Did you observe any evidence of other living things on the site? Did you notice any webs, nests, pathways, or even animals?
   - How does water move on the site? Are there any puddles? Any evidence of erosion?
   - What kinds of litter did you find? What might the location of the litter tell you about traffic patterns or wind direction?

   Ask students to take a few minutes to record their own questions about the site.

7. Reconvene the group and have them take one last look around the site. Leave the site as you found it and return indoors.

Consolidating the Data

Recommended Time: one or two class sessions

1. Take out the maps that students made from memory before visiting the site. Ask:

   - How do your new site maps reflect what you really observed rather than what you remembered? What are the differences between the two maps?

2. In small groups, have students compare their site maps. Ask them to look for similarities and differences, and to discuss discrepancies.

3. Then work together as a class to develop one large class map on the chart paper you have provided. Some of these logistics might help:

  - Draw in the outline of the buildings first.

  - Record the compass rose and the legend or key that students developed.

  - Then add the physical features.

  - Finally, sketch in the natural features.

  - Be sure to include any measurements students have recorded.

  - Make note of any signs of activity on the site that might indicate how it is used by different organisms.

4. Focus attention on the chart labeled *Questions* and invite students to share ideas, observations, and questions that arose as they worked at the site. Record their thoughts, and leave the chart posted so that students may refer back to it as they develop new projects. Mention again that questions are a valuable product of their experiences outdoors, and these ideas may become topics for further investigations.

Follow-Up: Using the Map in Upcoming Projects

Leave the consolidated map on display. Students may want to add to it, make corrections, and use it as a reference tool as they work on other projects, such as a field guide, a collection, or an animal behavior study.

The map is an important piece of the database. On it, students have noted the relative positions and approximate sizes of human-made features and natural organisms. They are increasingly aware of what resources exist at the site, and now know a great deal more about where the resources exist relative to each other.

The questions that students developed during the course of the mapping project should provide them with rich material to launch a series of more long-term and more individually tailored investigations. Students may have become interested in a particular organism or a particular relationship, and are motivated to carry out a more in-depth study that revolves around finding answers to their own questions.

Something Extra for the Teacher

Veteran teacher Allan Hayes relates the story of how his class mapped the schoolyard by using an aerial view of the site that they then divided up into zones by placing a grid over it. Archaeologist Elizabeth S. Peña tells how careful mapping helps her make sense of an archaeological site and solve some of the mysteries that it hides.

What Happened to That Pigeon? Creating a Schoolyard Map

Alan Hayes
Teacher, Nichols School
Buffalo, New York

My students prepared to begin drawing trees and shrubs in the school's court-yard, an area called the Quad. They had their journals and colored pencils and were in a quiet reflective frame of mind. As we filed out my classroom door, one student came running down the hall exclaiming, "I just found a dead pi-geon outside the grammar teacher's door. There is blood!" The whole class practically ran down the hall to see the carcass before the maintenance men had a chance to remove it. On the spot we decided to start investigating: stu-dents immediately began drawing and jotting down notes in their journals, while others searched the area, finding feathers scattered near the door and some under a nearby tree. "Look!" said one, and everyone glanced up to see a hawk perched high on a branch, peering down at us.

Back in the classroom, the students worked to sort their observations from inferences. They filled the board. When I asked what evidence they had that the two sightings were related, the class proceeded to carry out a crime scene–like investigation of what they had discovered. They gathered all of their ob-servations, including detailed notes about the pigeon's remains as well as eyewitness accounts from other students and staff. As the class put together a story of what might have gone on between the pigeon and the hawk, it be-came apparent that we needed a map of the Quad to be able to accurately re-construct the events and depict the movements of the birds.

I searched the Web for aerial views of our city. TerraServer was my first source and then a colleague showed me New York State's GIS site. From it I downloaded two images of the Quad, one at 0.06 miles across and the other at 0.12 miles. I copied these for the students, stapling the narrower view over the wider shot and asking them to figure out what was shown in the first picture.

Looking only at the narrow view, they mistook sidewalks for streets and a circle around a flagpole for a monument in a traffic circle. When they saw the wider view that included the school buildings, they instantly recognized ev-erything and were intrigued to see how their school and its surroundings looked from the air. Capitalizing on their interest, I had them use the scale provided (0.06 miles) to figure out how many feet across the image repre-sented. We had been doing decimal numbers in math, so this was a nice as-sessment opportunity.

The next day I had the students go outside to measure everything, includ-ing the heights of objects. Teams used tape measures and a trundle wheel. Altogether the teams logged over fifty measurements. We transferred their data to a transparency of the courtyard. One team developed a map from the com-bined data. I raised the question, "Is there a way we can organize our map, the aerial view of the Quad, so that anyone would be able to pinpoint where something is?" We discussed ideas for dividing up the space. The students were familiar with maps that use letters along one axis and numbers along the other. They agreed that positioning a grid over the aerial view would work, but they were uncertain what scale to make their map and how to go about laying out the grid. We had already taken three days to develop a map, so I decided to accelerate the process by giving the students an aerial view with a grid in place. Along with this new map with the superimposed grid, I gave the students three tasks: develop a measurement key where one box equals *x* dimension; draw a compass rose; and label the map.

The students are involved in the development of this map because they understand the value of having an accurate plan of the Quad. Already they are talking about plotting the hawk's movements through the area, wanting to see whether its presence affects the pigeons' movements. I see these investigations growing and changing through the year, with lots of opportunities for cross-curricular integration.

Mapping the Site to Answer Questions About the Past

Elizabeth S. Peña, Ph.D.
Director, Art Conservation Department, Buffalo State College
Archaeologist, Old Fort Niagara State Historic Site

Archaeology is all about context: Exactly where on the site did the pot sherd come from? Which soil layer contained the fragment of window glass? Which artifacts were found together? Not only does careful mapping help us make sense of an archaeological site, it also helps us solve archaeological mysteries.

Located at the junction of Lake Ontario and the Niagara River, Fort Niagara marked the Great Lakes entrance to the North American continent and controlled access around Niagara Falls throughout the colonial era. Now a National Landmark, the fort was built by the French in the late seventeenth and early eighteenth centuries, then taken over by the British in 1759. It became American some years after the Revolutionary War. Because of the fort's complex history and the wealth of historic maps available, our mapping process must include the standing structures on the site as well as the buildings that no longer exist. Using GIS (geographic information systems) allows us to digitize historic maps and create overlays with the current landscape, which we map using a laser transit. The data from the transit are downloaded and used to create plans that can easily be modified as work proceeds. Through GIS, we can access information about current and past landscapes. This information is extremely important in the creation of archaeological research designs and subsequent decisions about exactly where to excavate. Even with this abundance of information, however, we encounter plenty of surprises.

One field season, our research plan centered on the 1768 guardhouses, which had been razed by the end of the eighteenth century. The historical records and maps told us that one guardhouse was occupied by enlisted men, while the other was the officers' building. We were interested in status differences that would be reflected in material culture. While we did recover sections of the guardhouse foundations and associated artifacts, we also came across a large pit, filled with early nineteenth-century material.

This pit was a mystery. Its contents included several very large (and heavy!) iron kettles and a skillet, padlocks and door hinges, an intact glass bottle and many glass fragments, pottery sherds, animal bones, and military items such as gunflints and buttons. An examination of the artifacts gave us an early nineteenth-century date, based on the presence of brown transfer printed pearlware, whose production began in 1809, and oval-shaped copper alloy belt plates from the War of 1812 period. Plotting the pit's location on our GIS, we could see that it had been located adjacent to the "Red Barracks," the site of a short battle between American and British forces during the War of 1812. It seemed likely that the pit was dug and filled by the British when they took control of the fort, or by the Americans when they returned.

Whether the pit was created by the British or the Americans, some important questions remained. From the many historic maps of Fort Niagara, we knew that the parade ground had been a crowded place throughout the eighteenth century, with many small buildings clustered around the fort's large stone structure. By the nineteenth century, however, those small buildings had been torn down, and prevailing notions in military design included a large, grassy parade ground. Fort Niagara's nineteenth-century maps show that this was the case here, too. Why would the soldiers dig a big pit for garbage in the middle of the empty, grass-covered parade ground when they easily could have thrown the garbage into Lake Ontario or the Niagara River? Why would they mar the appearance of the ceremonial parade ground?

To answer this question, we turned back to our mapping project. This time, we focused on the palisaded, or fenced in, areas of the parade ground. When the outlines of the palisades were added to the GIS, we could see that the "Red Barracks" was very close to a fence line. This meant that the pit was not in the middle of the parade ground after all—it was in an alley! The pit was not visible from the parade ground, and did not disrupt its clean appearance.

At Fort Niagara, mapping the archaeological features, the standing structures, and the historic landscape allows us to draw clues together to gain an understanding of those who lived at the Fort in the past. The War of 1812 soldiers were not damaging the parade ground when they dug the pit; rather, they were disposing of unwanted items in a back alley, unseen from the flat, clean parade ground. While many more questions remain, thorough mapping allows us to understand the soldiers' actions and gives us insight into the past.

6

Creating a Field Guide

What Is a Field Guide?

A field guide is a useful tool whose main purpose is to aid in the identification of specimens in their natural environment. There are many excellent ones available both in book form and on the Internet (see Appendix B: Resources and References for a suggested list). Commercially published guides are often specialized and deal with a specified geographic region as well as a particular category of organisms such as plants, animals, rocks, and minerals. For example, there are field guides to the butterflies of North America, the flowers of the Rocky Mountains, and the shore birds of Florida.

A good field guide is organized as a visual key and uses colored drawings or photos of the specimens to aid in identification. The written notes usually include measurements; prominent distinguishing features, such as shape, texture, and color; range and habitat; and interesting facts about such things as life cycle, food preferences, or distinctive behaviors.

Why Create a Field Guide to Your Site?

In Chapter 3, Getting Started with Students, students were introduced to field guides, and probably have continued to use them to identify specimens. Now, they will look at field guides in a different way, to analyze their component parts and decide on how to format their own guide.

Students are on their way to becoming experts about the plants and animals at their study sites. One way to build on their growing store of data and to keep curiosity and investment in the site high is to have them create a field guide that specializes in your own corner of the world.

Students will have already indicated the locations of many organisms on the class map and taken notes about them in their journals. (See Figure 6–1.)

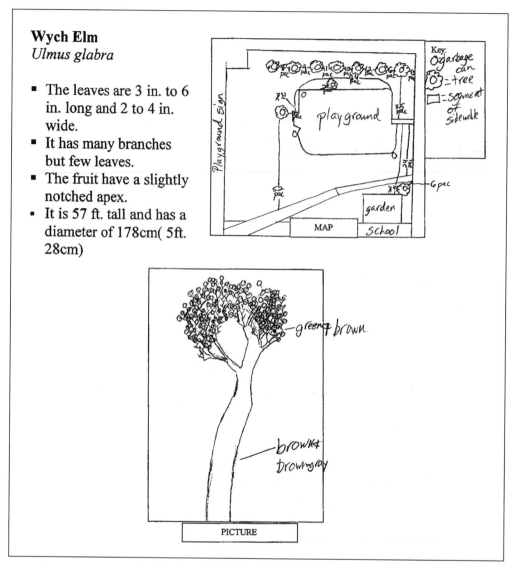

Wych Elm
Ulmus glabra

- The leaves are 3 in. to 6 in. long and 2 to 4 in. wide.
- It has many branches but few leaves.
- The fruit have a slightly notched apex.
- It is 57 ft. tall and has a diameter of 178cm(5ft. 28cm)

Key:
O = garbage can
O = tree
□ = segment of sidewalk

Playground Sign

playground

garden

MAP

School

green + brown

brown +
brown gray

PICTURE

FIGURE 6–1. Students in this class first generated maps of the schoolyard, and then went on to develop a field guide.

Both the map and the journals serve as jumping-off places and provide initial, if sometimes sketchy, data. Nevertheless, the data belong to the students and are a legitimate and authentic reason for students to return to their own records. Plus, in reviewing their initial data, students often come to realize that their journal entries are vague or incomplete, and are thus motivated to return to the specimen for more information.

Another aspect of the field guide project is that students may elect to focus on an organism that is of special interest to them, thereby encouraging
them to develop a sense of ownership in the project. Note that some organisms are easier to study than others, but try not to discourage students from
taking on the more difficult ones. Plants, and in particular trees, make good
subjects for beginners. They are relatively large, stationary, have many visible
parts, and change slowly over time. Moving targets like birds, flying insects,
and active but shy mammals require more patience and more diligence, but
are well worth the effort.

Cross-curricular links abound in this project. Creating a field guide promotes using skills in science, math and measurement, geography and social
studies, language arts, drawing and photography, and research.

When to Begin

Students need to have a familiarity with the site and with techniques for working outdoors productively before they begin to create a field guide. They should
also have collected some preliminary data before they select a specimen for
further study and inclusion in the field guide. The preliminary data collection
process that is part of mapping the site usually serves to heighten student
awareness of local organisms and to raise questions about what the organisms
are and how they live. So the closer study that creating a field guide requires
is a logical next step.

This guide would focus on plants and animals that students have observed
firsthand and identified themselves. It works well as an ongoing project, so
that whenever students complete an identification they can add the specimen
to the field guide. The guide could also become a chronicle of organisms discovered at the site over time, and students might pass down their field guides
from one class to the next. (See Figure 6–2 for a student-generated field guide
entry.)

Field Guide Formats

A quick glance at any two or three field guides on the same topic will show
that, although their content is similar, their formats may be quite different.
It's up to you and your class to develop a format that works for you, but for
consistency, it's important that everyone use the same format for each entry.
Here are some suggestions:

- For each entry in the field guide, use two separate, facing pages. Keep
  the pages in a class loose-leaf binder so you can add pages each time
  students identify a new organism.

- Designate one page for the written data and the other page for the labeled drawings.

- The basic written data include the common and scientific names
  of the organism, a description, measurements, and contextual

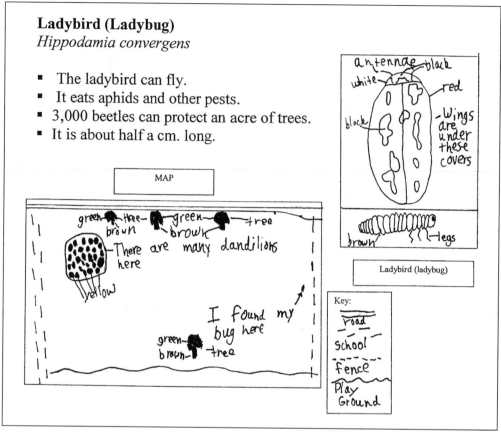

Ladybird (Ladybug)
Hippodamia convergens

- The ladybird can fly.
- It eats aphids and other pests.
- 3,000 beetles can protect an acre of trees.
- It is about half a cm. long.

FIGURE 6–2. This field guide entry includes drawings of the ladybird beetle in both the adult and larval stages, and indicates where it was found.

information such as where and when the organism was found, and by whom. To take the written description further, students might also include information from their research on such things as range and habitat, adaptations, behaviors, or other interesting tidbits. Figure 6–3 shows one of many possible formats.

- Drawings or photos might include side views, close-ups, and some of the surrounding habitat. Labels usually identify parts of the organism or point to significant identifying features. See Chapter 3, Figure 3–4 for an example.

- Use tabs in the notebook to identify categories of specimens. For example, make one tab for plants, another for animals, another for rocks, and so on. If the guide gets large enough, you might subdivide the categories. Within animals, for instance, you could add categories for birds, mammals, insects, and reptiles.

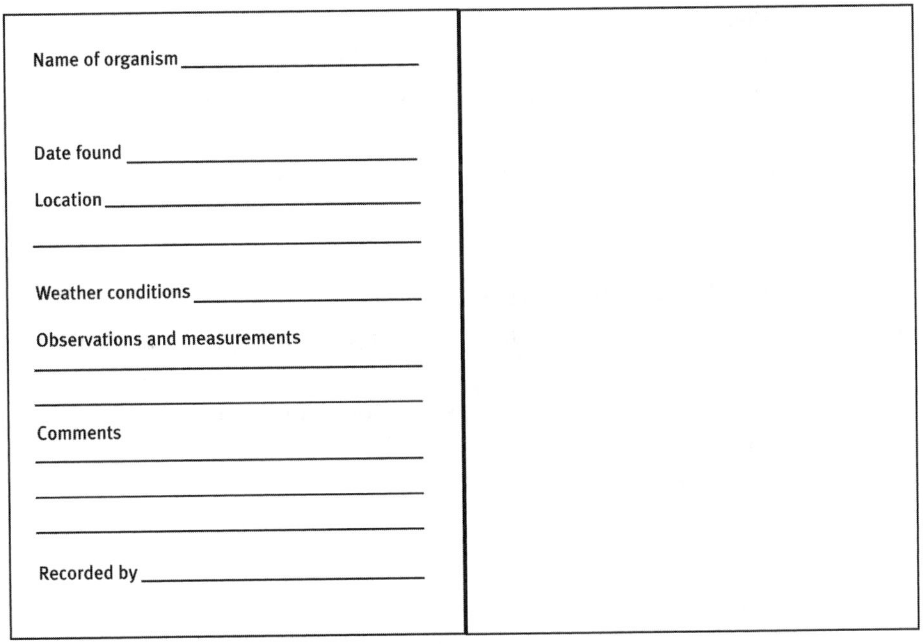

FIGURE 6–3.   A sample format for field guide pages

The following Investigation lays out a basic lesson plan for creating a field guide with your class.

The Investigation: Creating a Field Guide to the Study Site

The Lesson at a Glance

Students:

- analyze the elements of a variety of published field guides available in print and/or online
- develop a format to use for their own guide
- review their data to decide upon a specimen and a question about it to study further and include in the class field guide
- return to the specimen to collect more data
- identify the specimen
- create a line drawing with labels and a written description that includes measurements, contextual data, and common and scientific names, then post the entry in the class field guide
- add new questions to their journals and to the class list

Materials

- an assortment of commercially published field guides and/or access to the Internet (see Appendix B: Resources and References for a suggested list)
- one overhead master of How to Read a Field Guide (Figure 3–4, p. 33)
- an overhead projector
- two sheets of chart paper and markers
- one field journal per student
- one loose-leaf binder with lined data recording pages and blank drawing pages
- measuring tools, such as rulers, yardsticks, meter sticks, and tape measures
- hand lenses
- a compass
- thermometers
- colored pencils
- the class map of the site (see Chapter 5)
- a chart labeled *Questions*

Getting Ready

1. Gather an assortment of field guides on various topics. Make them available to students several days before you begin the project. If possible, also arrange for students to have access to some online field guides. Provide students with focus questions (see 1 below) referring to how the guides are organized and what kinds of information they present.

2. Make an overhead transparency of the page How to Read a Field Guide.

3. Label one sheet of chart paper *What's in a Field Guide* and the other *Our Field Guide Format.*

Preparing to Create a Field Guide

Recommended Time: one class session

1. Open a discussion on field guides and ask students what they have found out about the way that information is presented in them. Record their responses on the chart. To steer the discussion, use some of these prompts:

   - What are some of the things you noticed about the field guides?
   - How are they organized?

- What kinds of information do they give the reader? Let's list the different kinds of information found in a field guide.
- What is the purpose of the information the field guides give? How is the information meant to help the reader?

2. Focus attention on the overhead called How to Read a Field Guide and invite student comment on the types of information presented and the level of detail depicted. As students respond, add their comments to the chart.

3. Then, using the field guides, the overhead, and the chart of comments, ask students to work together in small groups to design a format for their own field guide. Encourage them to use a two-page spread, one for written data and one for labeled drawings. Let students know that the whole class will need to decide on just one format for everyone to use. That way each entry will contain the same kind of data and everyone will know how to read it.

4. Ask small groups to report on their format designs. They may sketch them out on the board or on a blank overhead transparency. Then ask the whole class to critique the designs and come to agreement on which elements they should use. Typically, the format designs will contain many of the same elements, so it is usually not difficult for most classes to come to consensus.

5. On the chart paper labeled *Our Field Guide Format*, record the final design. Leave it posted for student reference as they create their entries.

Creating an Entry for the Class Field Guide

Recommended Time:

- one class session indoors to decide on a study specimen and review data
- one class session outdoors to collect new data
- one or more class sessions indoors to refine and post the field guide entry

1. Ask students to review their journals and the class map of the site to search for questions about organisms that interested them. They may work individually or in small groups.

2. Once students have decided on an organism that they would like to study further, identify, and eventually include in the class field guide, ask them to compare the data they have already collected to the data they would need in order to make an entry in the field guide, following the format they have agreed on. In the majority of cases, this closer look at their own data will prompt students to go out and make more observations.

3. Arrange for another field session. Be sure to review the safety guidelines and the purpose of the field trip beforehand. Gather all equipment and take students to the study site. At the site, when students have begun to focus on more detailed observation, use some of these cues:

- What new data are you collecting? Why?
- What are you measuring?
- Have you recorded the new information in your journal? Are your drawings more detailed? Have you added labels?
- How will this information help you identify your specimen?

4. Back in the classroom, give students time to do the following:

- Consult field guides to make a positive identification of their specimen, if possible.
- Refine their drawings and written descriptions, and then record them in the agreed-on format. They will need to transcribe data from their journals onto the sheets of paper that will make up the field guide.
- Check with the teacher to make sure the entry is clear and complete.
- Post their entry in the class field guidebook.

5. Ask students to share the new questions that arose as they worked. Record the questions on the chart. Let students know that these questions are also an important product of their project, and may lead to further investigations.

Follow-Up: Taking It Further

Here are some tips about how to refine and make use of the class field guide:

- Give your field guide a title and put the title on the cover of the class book.
- Add a contributors' page, listing the names of all the students who have posted entries.
- Use the class map of the site (see Chapter 5) as an organizer and a way to locate each organism. Have students transcribe a smaller version of the map to include at the front of the field guide. Spread the map over two pages with numbered bubbles to indicate where each specimen was found outdoors and where data and drawings of that specimen can be found in the field guide.
- Put the field guide on display for others to enjoy. You might find space in the library/media center, a local business, a garden center, or nature center.
- Create an electronic version of the field guide by scanning it onto a CD or by posting it on the school website.

Students have created a lasting legacy, one that they can pass on to the class that follows them. At the end of the school year, you might turn the "passing of the field guide" into a brief ceremony in which the current class explains their work to next year's class, shares their list of unanswered questions, and encourages them to add to the field guide over the next school year.

Something Extra for the Teacher

Teacher Michael Milliman describes how students went from journal use and mapping to a need to produce a field guide for their own schoolyard. Wayne K. Gall, Ph.D., Regional Entomologist for the New York State Department of Health, chronicles the evolution of the field guide, an affordable resource dating from the 1930s that made the study of natural history accessible to a wide range of enthusiasts.

What's Out There? The Making of a Schoolyard Field Guide

Michael Milliman
Teacher, Amherst School District
Amherst, New York

I wanted an open-ended question to focus my students on their first outdoor investigation, one that would challenge them to look at their surroundings through naturalists' eyes. So, gesturing to the schoolyard, I began with the question, "What's out there?" Not surprisingly, the kids' immediate answer was "grass." Situated in a first-ring suburb, our school is typical of its kind, surrounded by a large, well-groomed lawn that is dotted with trees. At first glance it appears a pretty sterile place. "Is that all there is?" I asked. The students agreed that they'd need to do a more detailed survey to know. "We could write down what we see in our journals," Emma said. So we grabbed our field journals—the ones they had used to sketch the reproductive parts of flowers, describe mammal sign, and record many new questions—and headed outside.

Faced with the wide-open space of our school grounds, I wondered aloud how we could be sure that we would have a complete survey of the site. After a debate about the merits of various surveying and sampling techniques, the students agreed to divide the area into three zones. Each student was assigned to a zone, which they mapped in their field journals. I modeled some techniques for judging distance and for representing tree trunks versus tree canopies, but mostly I wanted the students to make a map that made sense *to them*.

Back inside after this initial outing, we compared everyone's maps and filled in some details. The discussion was lively. "Oh, that's right, there are three trees, not two!" "There was a flower growing by the playground benches? I didn't notice." After the class reached consensus, I asked, "What should we do next to find out what's really out there?" The general response was, "We need to go back outside! We need to look more closely."

I was pleased with their enthusiasm and the next day I sent them back to their zones to start looking for a living thing. I suggested that they search out an organism that they couldn't identify. Some kids caught insects. Some hunched down over weeds that the groundskeepers had missed. Some sat under a tree. Some even tried to squint at a bird flying from tree to tree, others at a squirrel chattering at them from a branch. Of course, some wandered aimlessly, but with a little guidance, they, too, began to find intriguing things emerge from the homogeneity of the grass. I think this was a key moment in their investigation. Now most of the students were hooked. They had found mysteries in the mundane, had detected the unknown and strange lurking in the known, and now they were primed to ask questions.

It was at this point that I realized that students were using the skills we had practiced in every science unit, and doing so independently. Driven by their curiosity, they were taking careful notes, incorporating measurements, and making detailed sketches from multiple points of view. They requested hand lenses, tape measures, and—most of all—answers.

Armed with their journal notes, students combed the library for information that would help them identify their specimens. We soon discovered the limitations of our otherwise well-stocked library's field guide section. The "tree sleuths" grinned, but everyone else grumbled. The sources we had were too basic. Our taxonomists were not satisfied to know they had spotted a beetle. Like all good detectives, they *wanted names*. Our elementary-level field guides just did not deliver. We were all sent scrambling to local libraries and our home bookshelves to find better resources. The students were on a mission, fueled by the idea that perhaps they just might have discovered a species of insect or weed heretofore unknown to science!

This convinced us: we needed a field guide of our own, so that if someone else stumbled on a bug, or a weed, or picked up a leaf, they would have a resource to use to identify the living things that make our schoolyard a home. The students set about designing the pages, discussing the information they would include to make this guide easily useable. Like many field guides, their pages had color illustrations showing defining characteristics, like leaf shape, coloration, and distinguishing marks. Students also chose to include a detailed map, locating where to find the plants identified, and where the animals had been spotted. Finally, each page listed information about the organism that they had researched, as well as measurements gathered from their investigations.

The students were very proud of the resulting field guide, especially when a copy of it was placed in the school's library, alongside the other natural history resources. I was really pleased to see how easily and well this project addressed so many learning goals in a thoroughly integrated way. Now I am looking forward to expanding this project each year with new classes. I can see that over time we will have a wonderful store of information that will allow us to explore new questions about our site. "Are the same things in our schoolyard today that were there three years ago? What's changed? Why?"

The experience of discovering something by themselves, for themselves, was empowering for my students. It taught them about the nature of science and themselves, as well as the schoolyard. What's out there? It turns out to be a lot more than grass!

Evolution of the Field Guide: From Species Identification to Community Ecology

Wayne K. Gall, Ph.D.
Regional Entomologist, New York State Department of Health

Recreational naturalists and professional scientists alike rightly consider the field guide to be fundamentally useful in identifying natural subjects or phenomena in the out-of-doors. Yet some may not realize how the focus of these compact, concise, and affordable field references has evolved over time from species recognition to interpretation of behavior, and ultimately, community ecology.

In the 1930s this new class of outdoor references, which focused initially on birds, satisfied a need that was not being met by the two- or three-volume scholarly regional treatises in ornithology available at the time, e.g., *The Birds of New York* by E. H. Eaton (1910, 1914), and *The Birds of Massachusetts (and Other New England States)* by E. H. Forbush (1925–29). Purchase of the latter was likely cost-prohibitive for the average naturalist, and regardless of cost, their sheer size and weight rendered them impractical for use in the field.

The year 1934 marked a revolution in the study of natural history. For it was then that the first edition of Roger Tory Peterson's *A Field Guide to Birds* was published, the forerunner of modern field guides and the flagship of the Peterson Field Guide Series. It ushered in a revolution not of new subject matter in natural history—wild birds were members of a pantheon of natural subjects already being studied by both avocational and professional naturalists—but rather a cost-effective, pragmatic approach to identification of birds (and later other natural subjects) in the field. By producing one pocket-sized, relatively affordable and comprehensive book that efficiently allowed the identification of the birds of eastern North America in the field, Peterson had solved the conundrum of high cost and lack of portability. Field guides were now accessible to the public at large, with the result that *A Field Guide to Birds* became an immediate best seller.

Then there was the advancement in *how* natural subjects were portrayed in Peterson's *A Field Guide to Birds*. This new approach consisted of juxtaposing original drawings of multiple species of birds on the same page, with arrows strategically deployed to focus attention on field marks used to discriminate similar species, or plumage differences related to developmental age and breeding cycles. This "Peterson System" of shorthand-illustration replaced the lengthy narrative descriptions employed in the more scholarly treatises. Although this technique was popularized and its use maximized by Peterson, it should be noted that the technique of employing arrows to draw attention to field marks of organisms was originated by Ernest Thompson Seton, who utilized them in the marginal illustrations of such classic works as his *Two Little Savages* (1903).

Thus the development of the field guide as an identification tool was consistent with the Chinese expression, "The beginning of knowledge is to get things by their right name." For the field naturalist, this expression embodies the fundamental importance and utilitarian use ascribed to the field guide in facilitating the identification of species, a basic function whose value cannot be overemphasized. And with Roger Tory Peterson's pioneering work, *A Field Guide to the Birds*, natural history became accessible to a wide cross section of the American public during the first half of the twentieth century. During approximately the next forty-plus years after Peterson's landmark work appeared, there ensued an expansion of topics in the Peterson Field Guide Series (e.g., Shells, Butterflies, Mammals, Rocks and Minerals, etc.), and a proliferation of field guide series from other publishers (e.g., the National Audubon Society Field Guide Series and the Golden Guide Series). Despite this expansion in the *scope* of field guides, the popular study of natural history did not *conceptually* progress beyond the identification of natural subjects.

A conceptual breakthrough in the study of natural history occurred with the appearance of the Stokes Nature Guides series, whose inaugural title was *A Guide to Nature in Winter* (1976), by Donald W. Stokes, with illustrations by Deborah Prince. The popularity of this nature guide fueled a relatively rapid succession of other titles in the Stokes series, including such titles as: *A Guide to Bird Behavior*, Volumes I–III (1979, 1985, and 1989); *A Guide to Observing Insect Lives* (1983); and *A Guide to Enjoying Wildflowers* (with Lillian Stokes, 1985). Note that the title of the Stokes series is *Nature* Guides rather than *Field* Guides. This difference is not merely one of semantics, for it reflects a shift in emphasis from identification to behavior, although certainly hints on identification of target species are sometimes given.

The key question posed by Stokes and his collaborators is one of fundamental importance to inquiry in natural history: once you know an organism's identity, what can you observe, or otherwise learn about it? The author of a Stokes Nature Guide assumes that the reader has previously identified a species, e.g., through the use of traditional field guides. The author then provides the reader with an in-depth look at the biology of selected species—including classification, derivation and meanings of names, folklore, life cycle, seasonal activities, how structure relates to ecological adaptations, habitat relationships, and especially their behavior.

The Stokes Nature Guides do an exemplary job of considering the behavior and ecology of targeted species, but generally only touch on their community ecology (the interaction of an organism with other organisms in its environment). An outstanding example of a field guide that takes the interpretation of an organism's community ecology to a new level of understanding is *Eastern Forests*, a Peterson Guide, by John C. Kricher, illustrated by Gordon Morrison (1988). This book is actually a field guide to ecology. Thus, rather than dwelling on identification of organisms as in traditional field guides, this guide focuses on interpretation of seemingly complex interactions in nature and recognition of ecological patterns.

Field guides have continued to evolve, offering readers more specialized knowledge and integrated approaches to viewing nature. But how do scientists today use field guides? Of course the answer to this question depends on the nature of the research problem being considered by the field biologist, his or her area of specialization, and taxonomic training or experience. As an example, let's take my own situation as an entomologist. When I conducted contract surveys for rare dragonflies and tiger beetles, the basic tools used in my fieldwork were a field guide to dragonflies and a field guide to tiger beetles. In conducting an annual butterfly count, I regularly referred to two field guides to butterflies. In conducting interpretive naturalist programs such as "Underwater Monsters," "Bugs by Nightlight," and an "All-Day Fern Walk," field guides to fish, aquatic invertebrates, moths, and ferns, respectively, were my most important tools.

Generally, scientists studying faunal or floral groups that are relatively less diverse, larger/showy, and better known such as birds, mammals, fish, amphibians and reptiles, trees and shrubs, etc., have greater success in using field guides to these faunal or floral groups in their field research. The corollary is that scientists studying relatively more diverse, smaller-sized, or less well-known faunal or floral groups, e.g., certain insects, grasses and sedges, mosses and liverworts, etc., find that field guides are less useful, are impractical to use in the field without magnification, or are simply unavailable.

7

Making a Collection

What Is a Collection?

A collection represents physical data, and as such is a record of what was present at a specific site at a specific point in time. The preserved specimens can be revisited and may be used as a way to compare and contrast specimens from other locations and other times. The real objects serve as a focal point for study, for making comparisons, and for exhibitions that educate the public.

A good collection is well documented. In fact, specimens that are not properly documented have very little scientific value. So once the fun of collecting is done, it is important to follow up with the harder work of identifying the specimen (with both scientific and common names), preserving it, and labeling it with contextual data such as the name of the collector, the date, the location, and observations on the habitat. See Figure 7–1 for an example of a specimen label.

Why Make a Collection?

We are all collectors to some extent, and most of us find satisfaction in having and handling real objects. Remember that coin or stamp collection, the box of buttons, or the shelf of figurines? The beauty of natural objects is that they are all free, just out there waiting to be found and to tell their story.

A collection is another vehicle for amassing data about the study site. It can also be a lasting legacy if one class passes its collection on to the next. Over the years, students may be able to detect changes in the site over time, and wonder about why these changes occurred—food for further research.

If you are planning a culminating event, the class collection is an obvious centerpiece. The objects themselves attract a great deal of interest and provide a unique and tangible overview of the site.

Scientific Name:	Common Name:
Location:	Collected by:
Date Collected:	Other Information:

FIGURE 7–1. A sample specimen label

There are countless cross-curricular links in this activity. Assembling a collection involves language arts, research, science-related skills such as identification and classification, and lots of good organization.

What to Collect: Dead or Alive?

Making a collection can be controversial and very upsetting to young students if it involves killing animals, even insects, so we recommend that your students take advantage of noncontroversial items at the site to collect. Plants and plant parts are a logical choice. Rocks and soil are also good subjects for study. If, however, students find already-dead arthropods, you may want to allow them in the collection. We would strongly discourage you from allowing students to bring in whole or parts of dead birds, mammals, reptiles, or fish for the obvious health reasons.

A short-term living collection would help satisfy students' needs to get a close-up look at some of the animals at the site. But students would have to be very careful to select only those animals (probably worms, slugs, snails, and arthropods) for which they could provide adequate accommodations for several days, and then return them to their natural habitat. In Figure 7–2 students use sweep nets to collect living specimens for a short-term collection.

Responsible Collecting

If you are going to collect, preserve, or maintain live specimens to investigate, it is important to be aware of proper procedures for carrying out the investigation in a responsible way, one that has minimal impact on the environment and ensures that the collection has scientific value. Here are some guidelines for collecting responsibly.

FIGURE 7–2.   Using nets to collect insects. Students will need to provide for the insects' short-term care, and then plan to release them back into their natural habitat.

Guidelines for Responsible Collecting

- **Research protected species.** State and federal government agencies publish lists of species that are protected by law. Please see their websites for lists of endangered and threatened species (vertebrate, invertebrate, and plant). If possible, consult a local expert to help you identify protected species in your area before you begin collecting.

    Be aware that bird feathers and nests are also protected by federal law and may not be collected without a special exemption. For more information see federal government websites.

- **Collect with a purpose and a plan.** Before you go outdoors, define your purpose for collecting specimens. Formulate a plan that lists the specimens you will collect and your reasons for collecting them.

    For example, if you are studying the trees on your site, you may decide to collect samples of leaves and twigs. The purpose might be to compare and contrast leaf types and leaf arrangements on the twig or to use the leaves and twigs to identify the trees. You might also decide to preserve the specimens and keep them in a permanent collection for future reference.

    It is even more important to have a purpose and a plan when collecting living organisms such as arthropods or growing plants. If you are investigating grasshoppers, for example, it will be important to identify the specific interactions, behaviors, or changes that you are going to observe, and it will be crucial to prepare carefully to provide for the needs of the organisms while they are in your care. Plan to return the living organisms to their original habitat as soon as the study is completed.

- **Collect only the specimens you need for the study.** Avoid the temptation to overcollect. Even a plant or animal that is common in your state may be considered rare in your particular location. Try to select the one best specimen for your purposes (Figure 7–3).

- **Record data to document your collection.** A collection only has value if it is well documented. Accumulate as much data as possible on your specimens. Include measurements, descriptions of color, texture, and aroma, and pertinent contextual data, such as the date, weather conditions, observations on habitat, and relationships to other organisms.

    Organize your collection of data so that you can retrieve it and make use of it. Develop a personal system of recording and numbering specimens in your journal. You may also create and make additions to databases online to keep track of your observations, illustrations, or digital photos.

FIGURE 7–3. Collecting a sampling of plants

The following lesson presents a basic plan for making a collection, and uses plants as the example. If you are planning to allow students to collect other kinds of organisms, you will find information on collecting and preserving them in most good field guides.

The Investigation: Making a Plant Collection

The Lesson at a Glance

Students:

- make a plan for collecting
- collect specimens from the site
- identify, preserve, and mount collected specimens, add labels and pertinent data

Materials

- one field journal per student
- plant labels
- scissors or pruning shears
- collecting bags and markers
- hand lenses
- rulers
- field guides to plants in your region
- several plant presses (see Getting Ready), either purchased commercially or student-made
- good-quality white paper, such as card stock, and white glue (for mounting specimens)
- copies or an overhead of Guidelines for Responsible Collecting
- the class map of the site (see Chapter 5)
- the class field guide of the site (see Chapter 6)

Getting Ready

1. Purchase or construct several plant presses. The press serves to flatten and dry out the specimens, which helps to preserve shape and color. It is possible to get good results by placing plant parts between the pages of discarded phone books, but it is also easy to construct plant presses that allow better air circulation and thus speed up the drying process. See Figure 7–4 for more detail. Here is what you will need:

   Materials for a Plant Press

   - corrugated cardboard all cut to the same size (about twelve inches square works well)
   - sheets of newspaper cut to the same size as the cardboard
   - two boards, one for the top and one for the bottom of the press

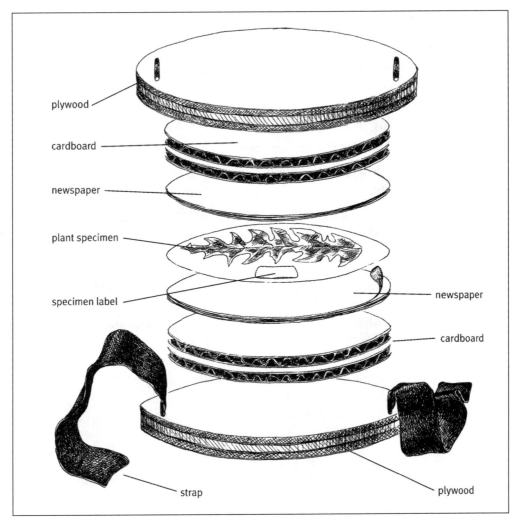

plywood

cardboard

newspaper

plant specimen

specimen label

newspaper

cardboard

strap

plywood

FIGURE 7–4. Component parts of a plant press

- strong twine, an old belt, or a bungee cord to hold the press together
- rocks, books, or bricks to weigh down the press

Microwave Drying

Another alternative is to dry the plants in a microwave oven. Place the plants between thick layers of newspaper, and weigh the "sandwich" down with a heavy ovenproof dish. Microwave for fifteen-second intervals. Check after each fifteen seconds so plants do not scorch.

2. If possible, plan for students to observe a professional collection of some type. For example, you may be able to take a field trip to a museum, herbarium, or nature center; if not, you may be able to view collections online. Or, you may find a member of the community willing to share a personal collection.

Preparing to Make a Collection

Recommended Time: one class session

1. Hold a class discussion on collecting. Use some of these prompts to guide the discussion:

   - What is the value of a collection? What can you learn from it? Why do museums, nature centers, and even private individuals keep collections?

   - What could we collect from our site that would be real, physical data? How would these specimens help to give a more complete picture of the site?

   - What questions do you have that might be answered by making a collection?

   - What responsibilities do collectors have? Distribute copies of Guidelines for Responsible Collecting or show it on an overhead; highlight the main points.

   - What equipment and tools would we need?

   - How will you preserve your collection?

2. Have students review their journals, the class map of the site (see Chapter 5), and their class field guide to the site (see Chapter 6) to search for and select the objects they will collect and the questions they will try to answer. (Please note that this activity is focused on plant collection, so you would have to modify plans if you decide to allow the class to collect other types of specimens.) When students have made their plant selections, ask:

   - Why do you want to collect that particular plant? What questions do you have about it?

   - How many plants or plant parts (such as leaves, blossoms, fruits or nuts, twigs, bark, roots) do you think you will need to give a good picture of the plant as it is right now? What might you add during another season of the year?

   - How will you preserve the plant?

   - How will you label it in the collection? What information should the label include? (See Figure 7–1.)

3. Gather the equipment suggested in the materials list.

Collecting Outdoors

Recommended Time: one or more class sessions

1. Do a quick review of safety rules, the purpose of the fieldwork, and the guidelines for collecting before leaving the building. Remind students that they will not only be collecting plants but will also be recording data in their journals, which they will later use to identify and label their specimens. Distribute supplies and equipment, and head outdoors.

   Note: Students may collect specimens in plastic bags and label the bags with their names. Or you may take the plant presses on the field trip and have students place specimens (labeled) directly in the presses. (See Figure 7–5.) The advantage of the latter method is that you can oversee what students are collecting, limit their number of specimens, and double-check labeling.

2. At the site, circulate among the teams and inject some of these questions:

   * What have you decided to collect? Why? How many parts of the plant will you take?
   * What are you recording in your journal? What are you measuring?
   * Do you have enough data to identify the plant? To make a complete label for it?

3. When students have finished collecting, return indoors. If they haven't already done so, have them place their plants in the presses.

FIGURE 7–5. Plants are placed directly into the press while out at the study site.

Identifying and Mounting the Collection

Recommended Time: one or more class sessions

1. Plants will take one or two weeks to dry. Be sure to check the presses regularly. In the meantime, have students use their journals to make positive identifications of the plants they collected and to prepare their labels as well as their descriptive pages. They may also begin planning the physical layout of their page.

2. When plants are dry, have students lay them carefully on a sheet of good quality paper (such as cardstock), one plant per sheet, and use dabs of white glue to attach the plants to the paper. Then have them add all pertinent data to the sheet. On a facing page, they may want to expand the entry to include more descriptive information in a narrative form.

3. You will want to keep the collection together in a way that allows classes to add to it. One way is to punch holes in the pages and place them in a binder. Another is to simply keep them as loose pages in a folder.

Follow-Up

1. Revisit the opening discussion to talk about the collection as another kind of database. Ask:

   • What is the value of our collection? How do the specimens help to create a more complete picture of the site?

   • What did you learn by making a collection? What new information and new data resulted?

   • What kinds of questions did the collection answer for you? What new questions arose?

   • What kinds of data would you need to collect to answer the new questions? What method would you use to collect the data?

2. Use the collection as a reference tool, both with your current class and with classes to come. It represents a picture of your site at a particular moment in time. Students in coming years can use the collection to compare and contrast numbers and kinds of specimens, look for patterns, and discern changes over time.

3. Put the collection on display. You may be able to make arrangements with the school media center, a local business, a museum, or a nature center.

Something Extra for the Teacher

Read as Bill Rogers, a teacher and naturalist, explains his method of collecting using a "rocket stomper" to select random locations for taking plant samples. Richard S. Laub, curator of geology at the Buffalo Museum of Science, talks about how a well-maintained collection may benefit generations of scientists to come.

Blasting Off to Scientific Collections

Bill Rogers
Director of Out-of-School Programs, First Hand Learning, Inc.

The third and fourth graders attending an after-school program at the Valley Community Center are always ready to explore outdoors. The Center has a great expanse of lawn where the kids often play games and participate in other recreational activities such as flying kites. I know when we go outside they will be immediately thinking sports, not nature, so I've come up with ways to meet their expectations and then transform them. For instance, when I want the students to begin to take notice of the plants in the lawn I use two stomp rockets as our tools for selecting plant-sampling locations. This unconventional collecting device consists of a simple plastic pump that when stepped on pushes an impulse of air through a hose to send a toy rocket soaring in flight. The students are excited to use the rockets and before we head outdoors we discuss how stomp rockets work and raise the question of the effect of launch angle on rocket flight. The students are ready to talk with authority about factors such as the weight of the stomper and the use of one foot or two. When they begin to use them outdoors, they identify variables such as wind speed and direction as well. The investigations are already underway.

I put the students in two teams, each having its own stomp rocket. Each student is to launch a single rocket and to collect one new kind of plant at the landing site. Students crop the plant at ground level using small scissors and then slip it into that team's plant press (a phone book). I put an adhesive note with the student's name and the stomp number at each specimen page. We rarely run out of new plant finds on the outbound trip to the pond (about five hundred feet or fifteen to twenty stomps). The return leg is another story and offers important evidence about the ability of this survey technique to capture the full diversity of the lawn's plant community.

The following week we examine our plant collection. The students prepare a plant sheet (half of a file folder) and write the date and location (stomp number) of the collected specimen. I assist them with the transfer of their dried plant to a piece of folder stock (cut to fit easily into a gallon-capacity, zip-closure plastic bag). We spend the next twenty-five minutes drawing the plant specimens in our journals. Students who have collected moss get to use a 20× microscope. The rest use a five- or six-power hand lens to look closely. Hairs, holes, veins, margins, flowers, and an occasional critter (e.g., a flattened aphid) are drawn, counted, measured. As the students are doing this work, I go around and ask them to tell me what surprised them about their specimens. In doing so, I am seeking to open a discussion about what they want to do next. I prompt them to write their questions and ideas for further study in their journals.

One of the most common questions has been, "How will these plants' roots look? Will they be all the same or just as different as the tops?" (Students inherently know that digging and getting dirty is almost as much fun as operating a stomp rocket!) If that is where the interest lies, I ask the students to select one of their dried plants for a root study. Each student makes a flag from a bamboo skewer and masking tape to mark the plant he or she wants to dig. I take a pry bar or a garden fork to loosen the soil at each skewer and then the students take a piece of wood (driftwood branches from the bank of a nearby river) and sharpen it on the sidewalk. This is their digging tool. (If they make it too fine at the tip, it snaps off when they are digging. They learn quickly.) As each student digs to get all the roots attached to the stem, it often happens that they discover tiny animals in the soil. I keep a container with moist soil and plant material (green and decomposing) ready for these specimens. All organisms become class property, but the finder completes the journal entry. Students tap their plants against the digging stick to gently loosen the soil before putting the plant into a plastic bag with a zip closure.

If time permits, each student collects a second plant of the same kind, as natural history collections look for variations by collecting multiples of the same types. It is possible to keep these plants alive in the bag for a few days. The advantage of doing this is that the students often discover small insects and other invertebrates such as snails, mites, and millipedes on them. These critters are stimuli for new questions, such as, "Are they feeding on the roots, leaves, on each other, on soil particles?"

Students draw and describe the pattern of the roots. We look at length (depth) spread and thickness and discuss the strategies each root system embodies. We sometimes separate the roots from the leaves (with plant shears) so that we can weigh each portion of the plant. Before we weigh the roots, we soak them in water to loosen any remaining soil. (A paintbrush can be used to dislodge the stubborn particles.) We damp dry the roots on paper towels before weighing them. The wet weight of the leaves and stem is compared to the root weight. Then we dry both portions (leaves and stems in a phone book, roots on a bed of newspaper) and reweigh. I ask the students to predict which portion will have lost the most water by weight. We use cardboard coin holders for close-up observations of roots, leaves, and stems. We staple (rather than tape) them closed so that they can continue to dry. The students use hand lenses at the beginning and then move on to low power (20× to 50×) microscopes to study the structures. They draw what they see and compare their observations with images that they find searching for *root cross section* on the Internet.

The options are almost endless. The class collection becomes a starting point for a myriad of investigations. I find that the group becomes invested in the materials they have personally gathered. The questions that arise can be very engaging because they result from a comparison of similar, yet unique, specimens—the collection. The work becomes involving, ongoing, changing, challenging, and rewarding, and it all began by shooting miniature rockets up into the air.

The Power of Collections: How a Scientific Database Can Both Instigate and Answer Questions

Richard S. Laub, Ph.D.
Curator of Geology, Buffalo Museum of Science

The first step in doing science is simply to observe, and for a student of nature one of the best ways to do this is to make a collection. The act of collecting is not, in itself, science. Rather, it equips an investigator with a tool for doing science. A collection constitutes a database, just like a census taken of the inhabitants of a city. It can be examined to answer a wide variety of questions, some of which may not even occur to investigators until many years after the collection was made.

Much of my career as a scientist has centered on the excavation of a site in western New York that contains an extremely rich assemblage of bones, plants, and archaeological materials. The age of these fossils and artifacts spans the period from the late Ice Age right up to the present, and they therefore constitute a record of the changing environments, animals, plants, and cultures of this region through the past thirteen thousand years. I and other scientists have made many discoveries, and written dozens of papers, based on what has been found at this site. The usefulness of this and other collections, however, extends well beyond the present, far into the future. Here are two ways that I have personally found this to be true:

First, one of the most important features of the scientific method is *reproducibility of results*. That is, other scientists must be able to replicate my investigation or experiment to satisfy themselves that my interpretations are correct. A well-designed collection, supported by good field data, allows others to examine the same objects and measurements that I encountered, so that they can see if they agree with my conclusions. If they do not, then there is reason to suspect that I was in error, and further investigation is needed to clear up the issue. Further, I have found that as my own fieldwork and collecting continued, I myself occasionally came to doubt some of my earlier interpretations. Revisiting the specimens and data allowed me to confirm or negate those doubts.

Second, I have been impressed that a seemingly unimportant specimen can hold significant information that does not become apparent until many years have gone by, and new questions have arisen. This illustrates an old saying: "Before the eye can see, the mind must first ask a question." In one example, a mastodon foot bone, collected in 1989, was placed in storage and hardly ever looked at. Eleven years later, the collection was examined by a paleopathologist, a scientist trained to recognize evidence of disease and injury in ancient bones. When this person saw the foot bone he was stunned, for it bore evidence that its owner had been infected with tuberculosis. This evidence spurred him to ask a question that hadn't previously been considered: "How common was tuberculosis among Ice Age mastodons?" He traveled to museums all over North America to examine as many mastodon skeletons as possible. And now his eyes were equipped to see what had been there for any-one to see all along. He concluded that virtually every late Ice Age mastodon in North America had been infected with tuberculosis. This one specimen opened to us new possibilities for investigating the mass extinction of large mammals that occurred at the end of the Ice Age.

In another example of late-blooming enlightenment, after some years we began to find peculiar "concretions," rocks formed by chemical processes, buried in the Ice Age layer at our site. Initially, these were regarded as nuisances, objects to be quickly removed in order to reach the ancient bones below. After a while, however, we decided it was prudent to collect some examples, if only to document their existence. After a few years, a geochemist (a scientist who studies the chemistry of rocks, soils, and ground water) asked to see one of these concretions. The rather undistinguished-looking object proved to be a Rosetta Stone, its chemistry providing us with a completely new understand-ing of the site. Initially, we had believed that animals and people came here to quench their thirst with water welling up through springs. Now we saw that our site had been a salt lick, where animals were drawn to drink water and eat soil that was saturated with minerals their bodies instinctively craved. This magnet for animals also drew people who were interested in those animals for food and other resources.

A collection, then, is the tangible representation of what a scientist ob-served in the field, and then interpreted. Tucked away as a relic, it is dead and useless. However, cared for by people who understand and respect its poten-tial, and visited by investigators who know how to ask questions of it, a col-lection becomes a living entity with the power to benefit many generations to come.

8

Observing Animal Behavior

What Kinds of Behaviors Might You Observe?

Animal behavior is a highly captivating subject for most students. But in order to carry out an animal behavior study, they will need to think more precisely about what constitutes observable behavior.

Animal behaviors fall into a number of distinct categories. Some major classes of animal behavior include construction, investigation, feeding, locomotion, reproduction, comfort, and aggression. Each class of behavior may be further subdivided into separate, discrete activities. For example, under locomotion you might observe walking, running, climbing, and jumping. To get down to an even finer level of distinction, the scientist may then describe the elements that make up each activity. So if you were to focus only on walking and jumping, for example, you might describe the relative speed and the number of feet on the ground at one time during each behavior.

In order to gather behavioral data that can be accurately analyzed and shared with others, scientists find it important to be very precise when defining specific behaviors. For an example, see Figure 8–1 for a list describing some of the behaviors of a common animal, the eastern gray squirrel. From this list of behaviors, students then developed a record-keeping chart. See Figure 8–2.

Pitfalls

The study of animal behavior is challenging for young learners. It is very tempting to anthropomorphize, and to ascribe human emotions or motivations to animals. We fall into these traps quite naturally, and begin to make assumptions that may not be accurate. For example, it is typical for students to assume that the monkey in the zoo is glad to see them and is smiling, when in reality

Active Behaviors

Foraging: Searching for food in trees, grass, at bird feeders

Feeding: Biting, chewing, and swallowing vegetation

Storing: Hiding food

Mouth wiping: Moving forearm across mouth

Jumping: Moving with all four feet off the ground, animal is airborne

Chasing: One animal in pursuit of another

Tail flicking: Movement of the tail not related to natural locomotion

Inactive or Resting Behaviors

Sleeping: Lying still, eyes closed

Self-grooming: Licking fur of own body while lying down

Sitting: Staying in one location in an upright position

FIGURE 8–1.   Some behaviors of the eastern gray squirrel

it is baring its teeth; or that the bird is singing because it is happy, when it is actually vociferously defending its territory.

In other words, the science of animal behavior is literally about describing what the mouth of the monkey was doing—lips pulled back, teeth exposed—and not about why it was doing it. We can't ask the animal if it was happy, or even if the target of the behavior was the student. At this stage we can't make assumptions about the animal's state of mind or motive. What's important is to record the behaviors themselves, and then look for patterns. Once a set of behavioral patterns is analyzed, the scientist may then begin to look for other connections that might reveal motives for the behaviors, but that is a task that comes much later and only after much more observation.

What Is Ethology?

Ethology is a branch of science devoted to the study of animal behavior. It deals with the objective study of animal behavior, especially (but not necessarily) under natural conditions. Good animal studies can also take place at the zoo, in the classroom, or even in your own home.

An ethologist uses a data collection instrument called an ethogram, which can be thought of as a catalogue of the separate and distinct behaviors that a specific organism displays. When developing an ethogram, the first task is to

Title:

	Sleep	Groom self	Tail flick	Sit	Forage	Feed	Store	Mouth wiping	Jump	Chase
Squirrel A										
Squirrel B										
Squirrel C										
Squirrel D										
Squirrel E										

Comments:

Date: Time: Temperature: Weather: Recorder:

FIGURE 8–2. A chart designed to record the observed behaviors of eastern gray squirrels

observe the animal closely and then to make a very detailed list of the behaviors the animal displays.

An ethogram can be either specific or general. For instance, you may decide to study just one type of behavior, such as aggression in blue jays or courting behavior in male guppies. Or you might try to describe all the different aspects of an animal's behavior and catalogue any and all behaviors that you observe. In this instance, you would try to answer a broad question, such as "What does a squirrel do all day?"

When observing the animal, the scientist uses the list, the ethogram, to check off the behaviors observed and the amount of time the animal is engaged in the behavior. A completed ethogram, then, may be thought of as kind of a time map or time budget. It could be used to answer questions relating to what percentage of time the animal spends on different activities such as foraging for food, climbing, or sleeping; how long the animal spends in social encounters; or how many times an hour the animal grooms itself.

In all cases, the first step is to observe and then describe a specific behavior in very precise language. For example, one description from a scientist observing a dog and its trainer might read, "The dog moves directly toward the trainer while looking at the trainer." Then the scientist would assign a name to that behavior, such as "approach." In further observation sessions, the scientist would tally the number of times the animal approached and when. But the scientist would not try to interpret the behavior or say why the animal approached. Therefore, there would be no speculation about the possibility that the animal was looking for a treat or liked its trainer—only an indication that it did display the approaching behavior so many times within a specific time period.

The scientist would no doubt have a long list of behaviors to catalogue during the course of the one-hour dog-training session. These might include ten, twenty, or even more categories, such as head in the air sniffing, nose to the ground sniffing, whining, yelping, tail wagging, and pointing. For students, however, it works better if they begin with just one or two behaviors, and then grow the list gradually as they make more observations.

How to Grow an Ethogram

If broken down into smaller pieces, students can successfully develop a simple yet quite respectable ethogram. It all begins with observation, so as a beginning exercise, we suggest that students focus first on three big questions as they observe their subjects:

- What did you observe the animal do?
- Where did the animal do it?
- Were others of the same kind of animal involved?

Then, after they have made their preliminary observations and identified and described one or two behaviors in context, students can go on to draw up a

Data Sheet

Species: Agouti

Observer:

Date:

Time:

Conditions:

Category of Behavior

Groom— self	Walk	Eat	Drink	Gnaw	Dig

FIGURE 8–3. Behavioral data sheet of the agouti

simple data sheet, whether you call it an ethogram or not. In Figure 8–3, an animal behaviorist at a zoo studying the agouti, a large South American rodent related to the guinea pig, has selected six simple behaviors to observe and record. The observer would make one tick under the appropriate heading each time the behavior was observed during the specified time period.

Students may find that a close observation of one behavior leads to the discovery of another one. In the case of the agouti, for example, the observer might notice that the animal spends some time sniffing the air. This may prompt the observer to expand the data sheet to include information on the new behavior. Thus, the record-keeping strategy may grow in small and manageable increments and eventually include data on more and more behaviors.

Why Study Animal Behavior?

Animal behavior is a naturally fascinating topic and one that piques student interest. It can become a highly motivating project that builds skills at a new level. In the course of studying animal behavior, students will be engaged in close, long-term observations. They will, of necessity, have to work in cooperative teams in order to gather and analyze data. Things usually happen too fast for one person to take in and record.

An animal behavior study of this type takes students several steps further on the road to inquiry. It asks them to formulate their own questions, develop their own tool (the ethogram or another, simpler record-keeping chart), keep scrupulous records, look for patterns, and come to conclusions based on their own data. There is also a very strong language and literacy component not to be overlooked. In generating the descriptors, students are forced to be very precise and very detailed in their use of language. It is not enough to say that the dog wagged his tail. A vivid set of descriptors might include the direction of the tail movement, the speed of the back-and-forth motion, and the length of time the behavior took place.

There are a number of websites that publish ethograms. Do an online search for some simple examples to present to the class.

Adapting Examples of Ethograms

Many examples of ethograms are available. We have provided several, and there are many more online. Encourage students to use the examples as springboards to developing their own record-keeping sheets, whether you choose to call them ethograms or not.

An Opportunity to Use Technology

In addition to acquiring and using new skills, students may be encouraged to use more technology to gather and record data. It is a real challenge to keep track of everything an animal does in a particular setting and in a particular time frame. One very useful tool is the video camera. Students can play back the tape frame by frame to examine and record notes about animal behaviors, and multiple observers can view the same set of behaviors. Audiotape recorders are also useful, and data can later be transcribed in journals. Computers will be a big help in storing large quantities of data, in graphing results, and in displaying conclusions.

When to Begin: What Skills and Experiences Will Students Need?

Recognize from the outset that developing and conducting animal behavior studies will be a long-term and very engrossing project, and that to carry out the study effectively, students will need a base of skills and experience on which to build. If they have already mapped the site, created field guides, and made collections, they will have amassed many of the prerequisites, but will need to organize and use their skills in a whole new way.

Students' skills in observation have most likely become more and more acute with practice. Now, however, they will need to transfer that skill to observing objects in motion. They will also need to devise a system of how to work within their team, perhaps assigning some to be watchers while others record in journals or film or record the activities on an audiotape. This kind of teamwork requires a high level of organization and an attitude that fosters cooperation.

Selecting Questions to Investigate

Out of student observations in previous parts of the outdoor experience will have come many questions, and ideally, there is a class chart that provides a cumulative record of these. Now it is time to sift the questions, find ones that pertain to behavior, and analyze those questions to figure out if they involve distinct behaviors that can be observed and quantified in some way. This also requires a rather high level of skill and much guidance by the teacher. For more information on identifying investigable questions, please see Chapter 9.

Inasmuch as it is possible, the record-keeping chart or ethogram should be of the students' own design. It will be important to provide examples of record-keeping strategies and even more important to help students define the exact behaviors they will study. Data collection may take quite some time, and students will need to persevere, make repeated observations, and tally the behaviors systematically. Finally, they will have to organize and analyze their own data and try to make sense of their findings. Did they answer their question? What data can they show to support their conclusion? Students then make public presentations and share their research methods as well as their findings.

Outdoors or Indoors?

Preferably, ethograms serve to record animal behavior in the animal's natural setting. If you are fortunate enough to have a site with a population of wildlife, such as squirrels, rabbits, insects, or birds, these would make excellent outdoor subjects.

But there are also other wonderful possibilities indoors. You might have a class aquarium. Lay a transparent grid over the outside to define zones, and then record where the catfish spends most of its time or which fish the angel fish chases and how far. Collect soil creatures outside, such as crickets, pill bugs, or worms, and build them a terrarium that replicates their natural conditions so you can study their reactions to light or moisture. (See Figure 8–4.) Set up a bird feeder outside the classroom window and observe what times different types of birds come to it or how different species of birds behave. Take a trip to the zoo or the nature center and find out what behaviors

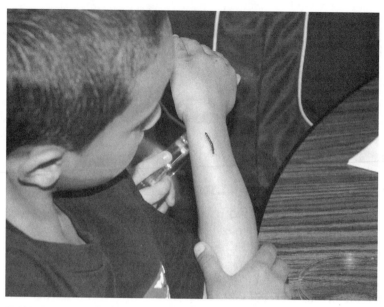

FIGURE 8–4.   This student has brought a millipede indoors to study
its locomotion.

captive animals exhibit. The possibilities are limited only by your students'
imaginations.

The following lessons lay out a basic plan for developing and using a
record-keeping strategy such as a chart or an ethogram to study animal
behavior.

The Investigation: Planning and Implementing a Study of Animal Behavior

The Lesson at a Glance

Students:

- define what we mean by animal behavior
- review their own questions and highlight those that pertain to animal behavior
- select an animal and a question for further study
- develop a preliminary plan for research
- observe the animal subjects closely
- refine their research plan and devise a record-keeping chart or ethogram
- collect data
- analyze their data; draw conclusions from the data
- present their findings

Materials

- the chart of questions that students have generated in previous activities
- chart paper and markers
- one field journal per student
- recommended: a video camera; an audiotape recorder; access to the Internet; computer programs for storing data, creating graphs and charts, and producing final reports
- for captive animals: a terrarium, aquarium, or other suitable container in which students set up a habitat to safely house the animals for a reasonable time

Getting Ready

1. Hang the charts of questions that students have generated in previous lessons.
2. Gather the materials recommended in the list above.

What Is Animal Behavior?

Recommended Time: one class session

1. Explain to students that in order to carry out a study on animal behavior, it will be useful to think first about what exactly we understand by behavior when we observe it in animals. Ask students to watch you carefully while you perform a simple "animal" behavior.

   - Wave your arm back and forth, bending it at the elbow, with the palm open and facing the class. Repeat three times.
   - Ask students to describe the behavior. Typically, they will say you waved. That is, they will name the behavior rather than describe it. Tell them that the single word *wave* is not a detailed description of the behavior. In fact, it could mean something completely different when it refers to waves in the water. How could we do better? Let's try again. This time try to notice which body parts are involved, what they do, how many times, and for how long.
   - Repeat your behavior. This time press students to describe your action in step-by-step detail, breaking it down into component parts, naming the body parts involved, estimating the duration of time, and noting how many times the behavior is repeated.
   - Record their descriptors on a chart.
   - Now, following the recorded description of the behavior, have students act it out. Does it work? Is it exactly the same behavior as they watched? This is a good test of how accurate the description of the behavior really is.

2. Practice once again. Ask students to watch the new behavior.

  • Give a big, wide smile, with lips drawn back and teeth exposed. Make it last for about five seconds, and then repeat once.

  • If no one says, "You smiled," you have made progress. As before, prompt students to describe your behavior in a step-by-step fashion, again breaking the behavior down into its component parts, telling how long it lasted and how many times it was repeated.

  • Once again, record the descriptors, and then ask students to use the description to replicate the behavior.

3. Next, make the point that when describing animal behavior, we don't necessarily know the reason for it, so it is important to stick to the facts of the behavior and not what we think they mean. Ask:

  • What if you visited a zoo and there was a monkey who faced the crowd and did this? (Smile again.) What could you say about that behavior?

  • Hopefully, students will describe the monkey's behavior without interpreting it.

  • But if students say something like, "The monkey was happy to see me, so he smiled at me," there are two aspects of the behavior to discuss. One is that we have no way of knowing how the monkey is feeling or what he means by his behavior. In fact, in some animals, baring the teeth is a threatening type of behavior. Like a baby, the monkey can't tell us why he is performing this behavior. The other aspect to discuss is that we cannot assume that the behavior is directed at us. For example, a gassy baby seems to smile at anyone within range. So it's not about why but about what the animal is doing.

4. Practice again.

  • Move your whole arm from the shoulder in an up-and-down fashion. Say that you are still at the zoo, this time in the penguin area. What could you say about this penguin's behavior?

  • If students describe the penguin's behavior without interpreting it as a wave directed toward them, you have made the two points. If not, discuss again that we don't know what the animal means by its behavior or why he did it, or at whom it is directed.

5. Go back to the chart of descriptors that students recorded on the two behaviors you modeled (the wave and the smile). This time, say that once scientists have listed the component parts of a behavior,

they usually give the behavior a name, thus creating a kind of a dictionary. Ask students what they would like to name the two behaviors and accept a reasonable name that all can agree upon. The most obvious would simply be wave and smile.

Developing a Preliminary Plan to Study Animal Behavior

Recommended Time: two or more class sessions

1. Give students time to review the chart of questions they generated in previous activities as well as their own questions recorded in the journals. Ask them to select questions that pertain to animal behavior that might make good topics for further study.

2. With the class, go through the chart of questions and mark those that relate to animal behavior. Use another piece of chart paper, if necessary, to add more questions that students contribute from their journal notes. As you go through the questions, ask about each one:

   • Why do you think this would be a good question to study further?
   • Does it interest you?
   • Is there something about the behavior that you could measure or time? What kinds of quantities could you record as data?
   • Have you observed the animal enough to be confident that your question is doable, practical, and safe (both for you and for the animal)?
   • Is this an animal that you will be able to observe regularly? Is it one that we see often at our study site? Or would you have to go somewhere else to observe it?
   • Could you bring the animal indoors and place it in a habitat that is like the one it lives in outdoors? How would that make it easier to study? How might moving the animal indoors affect its behavior?

3. Note that students will need to go through one more important preliminary step before beginning their study: They will need an extra period or two to observe their subjects to find out if the project is in fact doable, and then to pinpoint the exact behaviors they want to scrutinize. For example, if a team has decided to study squirrel behavior, they may need to spend some time just looking at squirrels to figure out which behaviors they really can observe. If students find that squirrels just aren't in the study area during the times they have available, they may have to change plans. In fact, during the preliminary observation period, several teams may decide to change or modify their original plans, and

this is fine. Have students focus on three questions as they make their observations:

- What did you observe the animal do?
- Where did the animal do it?
- Were others of the same kind of animal involved?

Encourage students to use their journals to note what the animal is doing and to add illustrations and new questions.

Now is also a good time to caution students again about attributing human behaviors or motives to animals. Remind them that it is important to be objective and to record only what they observe.

4. Explain to students that scientists who study animal behavior often use a tool called an ethogram, a kind of time map of behaviors they observe. Present students with several examples of ethograms such as the one in Figure 8–2 pertaining to the eastern gray squirrel or Figure 8–3 on the agouti, or have them look at some of those that are illustrated online. Discuss the ethograms with the class. Ask:

   - What is the purpose of an ethogram?
   - What kind of questions were the scientists trying to answer? What kind of data did they record?
   - What tools do you think they used?
   - Many ethologists work in cooperative teams. Why do you think that might be a good way for us to work? What are the different roles that team members might play?

5. Then ask the class to break into small groups of about three to five, and have the groups select one good question they would like to study about animal behavior. In their journals have them record the following:

   - the animal they will study
   - the animal behavior(s) they will observe and describe
   - the question(s) they will try to answer
   - their plan for conducting the study
   - the tools they would need

Planning is a crucial step, and students need sufficient time to develop their plans. Before students launch their behavioral studies, be sure to review each team's plan and give guidance where necessary.

Examples of Animal Behavior Study Topics

Here is a sampling of topics that students have found rich in potential. In each case, the students would first have to define the behaviors to be observed and then devise a way to count or measure them. In some cases, they might also set up experimental conditions, such as a terrarium that offers both dry and moist areas or dark and light areas. Some observations, such as those involving birds, squirrels, or flying insects, might benefit from the use of special equipment, such as binoculars or video cameras.

Indoor Topics for Investigation

- How many different ways can a worm (or mealworm, pill bug, cricket, ant) move? Do they usually travel in a straight line? Do they travel toward or away from dark areas? Do they travel toward or away from moist areas?

- Where in the aquarium does the catfish (or guppy, snail, goldfish) spend most of its time? What does it do on the bottom of the tank? How often does it swim to the top? How does its tail move?

- How many different behaviors does the classroom pet (rabbit, guinea pig, mouse) exhibit in one hour? Which behavior does it do most? Least?

At Home Topics for Investigation

- When are most people in my house sitting down (or reading, eating, talking)? For how long do they sit in one place? Does the time a person spends sitting down differ by age?

- What does my dog do when he is out in the yard by himself? (The observer would stay indoors at a window used as an observation post, and perhaps use binoculars.)

- How much time does my cat spend sleeping, grooming, eating, and walking around?

Outdoor Topics for Investigation

- How long does it take for an ant to find a donut? A piece of meat? A piece of fruit? What direction does it move in? How does its speed change? What does it do when it finds food?

- What does a butterfly do when it lands on a flower? How long does it spend on one flower?

- What does a bluejay do when other birds are at the feeder? How does it behave toward other birds?

- What do squirrels do all day? How do they spend most of their time?

- What different kinds of sounds do bird make? When do you hear the most bird sounds?

- At what time of day do we see rabbits on the lawn? What are they doing? For how long?

Refining the Plan and Developing a Record-Keeping Strategy

Recommended Time: one or more class sessions

1. Once students have selected an animal subject and a question for study, they need to further refine their research plan and design their record-keeping strategy, which may take the form of a chart or an ethogram.

   Mention to students that they may start very simply, with just one or two behaviors to observe. Then as they notice more behaviors, they could expand their record-keeping system to include the new behaviors. The chart or ethogram they design should have a list of the behaviors to be observed and spaces for tallying up those behaviors as they occur within a specific time period.

2. Give student teams ample time to design their record-keeping plans. Be sure to review them and provide guidance where necessary. Check that students have defined a list of behaviors and have then given the behaviors names. Then check that they have used the list of behavioral names to construct a chart on which they can tally their observations.

Implementing the Plan

Recommended Time: will vary

1. Set students to work on their projects. Check in often with each team and ask:

   * Is your plan working?
   * Have you been able to observe your animal regularly? How many times have you recorded data?
   * Are you able to keep complete and accurate records? How are you recording data? Are you happy with your record-keeping system? Is it too simple? Too complex? Do you need to modify it?
   * Do you need any additional tools or resources?
   * How much time do you think you will need to complete the study?
   * Does everyone on the team have a chance to contribute? How are you working together?
   * Are you finding answers to your questions? Are you finding more questions? Are your questions changing? Has anything been surprising to you?

2. Set a date for students to complete their behavioral studies.

Analyzing the Data

Recommended Time: one class session or more

1. When students feel they have completed their observations, ask them to look carefully at the data they have collected. Ask:

   - Review your question(s) and observations. What patterns of behavior did you observe that pertained to your question?
   - Did you find possible answers to your question(s)?
   - What data do you have to support your conclusions?
   - What new questions came up? How could you find answers to the new questions?

2. Give student teams time to organize their data and prepare to make brief presentations to the class, or to a wider audience such as another class or parents, if you choose. You may want to ask them to include the following information in their presentations:

   - the name of the animal and the question they were trying to answer
   - their methods of observation, the tools they used
   - the data they collected and how they recorded it (For clarity, the data may be presented on graphs or charts.)
   - their conclusions
   - new questions that arose

Presenting the Data

1. Invite each team to give a brief presentation of their study. Have presenters tell about their research methods, how they conducted the study, what kinds of tools and equipment they used, and what they found out. Encourage the class to ask questions of the presenters.
2. Take the opportunity to assess students during the presentations.

Follow-Up

1. Hold a discussion about the skills and processes involved in doing animal behavior studies. Ask:

   - What different skills did you have to use to carry out the animal behavior study?
   - What did you find most challenging about doing animal behavior studies? What was most valuable?

- What would you do differently the next time?
- What did you discover that was surprising, that you never knew before?

2. Animal behavior has captured the imagination of many fine scientists who have devoted lifetimes to studying creatures great and small. Interested students will enjoy learning more about their lives and work. Suggest that they find out about some of the following:

- Konrad Lorenz's imprinting studies with geese
- Karl von Frisch's work on deciphering the "language of bees"
- E. O. Wilson's creative and detailed research about ant behavior
- Dian Fossey's daily observations of gorilla groups
- Jane Goodall's forty-five-year-long observations of chimpanzees

3. Find an opportunity to display students' work publicly—either online, in the school media center, at a zoo or nature center, or in a local business such as a pet shop.

Something Extra for the Teacher

After-school volunteer and coauthor Kristen Gasser tells about doing a squirrel behavior study with seventh-grade students. Sara Morris, a professional ornithologist, recounts her experiences watching bird behavior, beginning with a memorable event in the fourth grade.

Squirrels Under Scrutiny: Using Ethograms to Track Behavior

Kristen Gasser
Director of Publications, First Hand Learning, Inc.

The kids in my after-school group had begun their observations of squirrels in the fall, when it became clear that we were lucky enough to have these urban mammals living in and around the trees outside our classroom windows. The students were most interested in "what the squirrels are doing." We began slowly, observing squirrels, writing about them in our field journals, and then discussing what we had observed. After these initial observations I decided to up the ante and introduce the idea of using an ethogram as a way to document behavior.

To gather useful data on squirrel activity I knew we would need to have very specific definitions of behavior. So I found an ethogram enumerating squirrel behaviors on the Web and we modified it to suit our needs. This provided the group with a starting point, a list of behaviors written in a straightforward, descriptive way, but they still needed to turn this list of behaviors into a usable format for recording behavior in the field. Each student came up with ideas for charts and tables, then they pooled their ideas and honed them, end-ing up with a basic format. From there the kids discussed how their recording sheets would differ depending on the questions they were investigating. How would you best keep track of two squirrels' interactions? What if you were curious about determining factors such as the amount of time spent doing certain behaviors? Could you design a recording format that would allow you to note the sequence in which squirrels performed specific actions? The challenges of designing recording tools made the discussions interesting and helped winnow the students' initial questions down to productive, investigable ones.

One group of students became focused on the movements squirrels made. They wanted to compare the time spent in motion with the time spent at rest. They came up with some interesting questions to investigate: What percentage of the observation time does the squirrel spend moving around? What percentage is spent at rest? How will this ratio average out over time? Does it vary between animals? If squirrels spend a high percentage of their time moving around, why do they expend their energy in this way? Does it help them survive? If so, how?

Another team soon became focused on the squirrels' food-gathering behaviors. The ethogram reproduced in Figure 3 is a record of observations made over a ten-minute period on a clear afternoon in early February when the temperature was twenty-eight degrees Fahrenheit. Each tick represents approximately ten seconds. The students were interested in documenting the amount of time spent on other activities (running, sitting, interacting with others, etc.) in an effort to answer the question, "What percentage of a squirrel's time is spent on food-related activities?"

The students have been getting more adept at using ethograms to chart the actions of our schoolyard squirrels, but it has been tough to catch all the behaviors. We've decided to use a video camera to document squirrel activity more accurately. As we continue this project, I'm interested to see what happens as the students amass more data and begin to try to analyze it.

Observing Bird Behavior

Sara R. Morris, Ph.D.
Professor of Biology, Canisius College

When I was in fourth grade, I remember watching an Eastern Bluebird outside my classroom. It was spring, and the male bluebird was attacking the window. I learned later that this was probably because the male could see his reflection in the window and was attempting to drive the intruder away from his territory. I have seen this behavior many times by many species of birds since that time, but I always remember that first observation. That was the beginning of my career as an ornithologist, a scientist who studies birds.

Observations of behaviors can help in the identification of particular birds. A bird creeping down a tree headfirst is likely to be a nuthatch. Birds creeping up tree trunks could be woodpeckers or Brown Creepers. These clues to identification can help make bird-watching experiences richer and

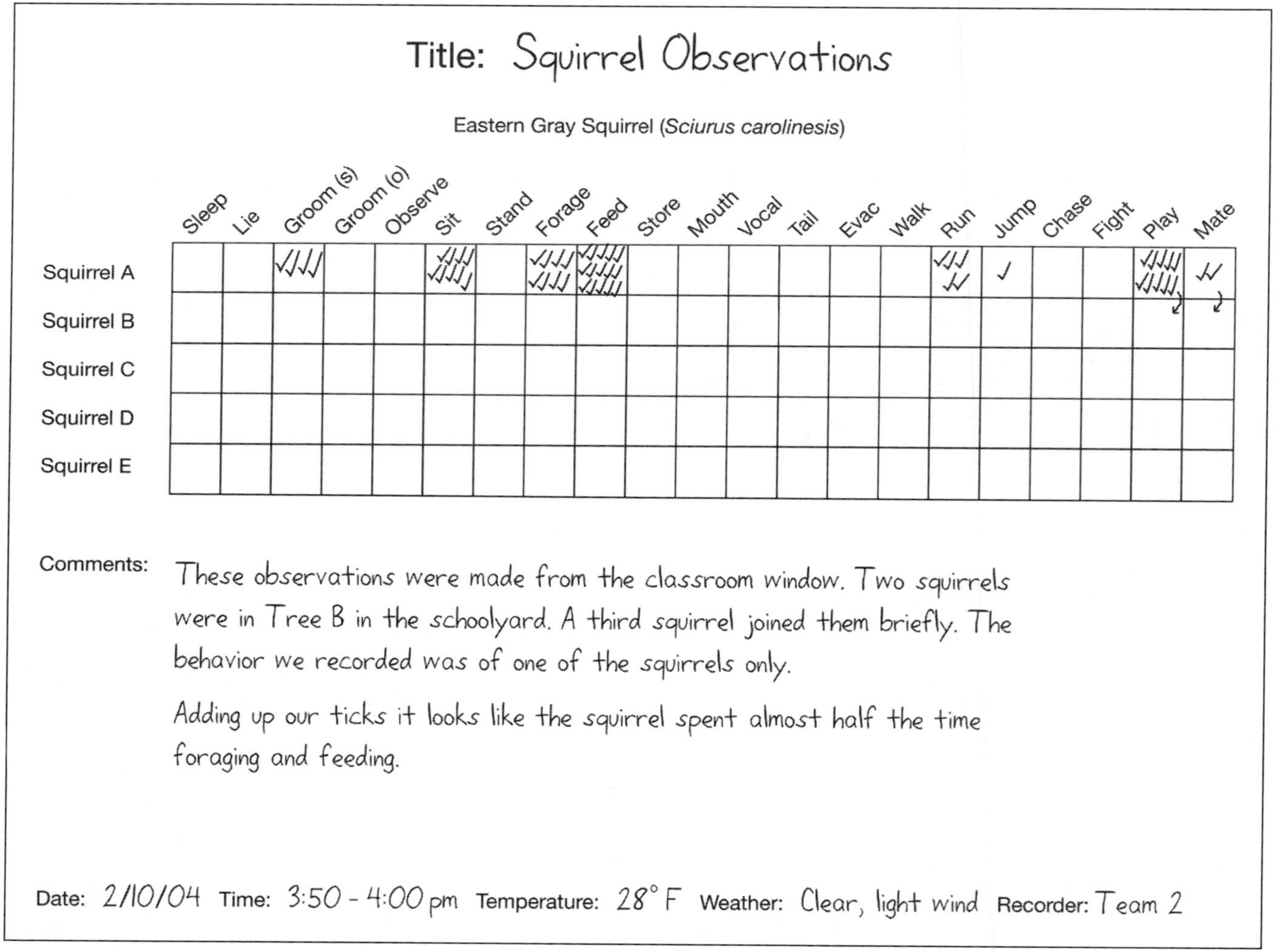

Title: Squirrel Observations

Eastern Gray Squirrel (*Sciurus carolinesis*)

	Sleep	Lie	Groom (s)	Groom (o)	Observe	Sit	Stand	Forage	Feed	Store	Mouth	Vocal	Tail	Evac	Walk	Run	Jump	Chase	Fight	Play	Mate
Squirrel A			////			卌		卌 卌	卌 卌							卌 //	/			卌 卌	//
Squirrel B																					
Squirrel C																					
Squirrel D																					
Squirrel E																					

Comments: These observations were made from the classroom window. Two squirrels were in Tree B in the schoolyard. A third squirrel joined them briefly. The behavior we recorded was of one of the squirrels only.

Adding up our ticks it looks like the squirrel spent almost half the time foraging and feeding.

Date: 2/10/04   Time: 3:50 – 4:00 pm   Temperature: 28° F   Weather: Clear, light wind   Recorder: Team 2

FIGURE 3.   A chart recording the behaviors of the eastern gray squirrel during a ten-minute interval

identification easier. My first observation of a Reddish Egret was in a salt marsh in Florida; I saw a bird "dancing." It dashed in one direction, twirled, and dashed back. The colors were right for a Reddish Egret, but the behavior made me even more comfortable with my identification.

Watching the behavior of familiar birds can also be rewarding by providing clues about their lives. When I see a Ruby-throated Hummingbird come to my bird feeder, I observe it for a while and am often rewarded by seeing it come to rest on its favorite perch. Then later in the day, I can look for the bird and often find it perched in that same spot. Similarly, watching what birds do when they are foraging can lead to exciting discoveries. If you see a bird carrying food, it generally means that the bird has chicks in a nest since birds ordinarily eat what they find as soon as they procure it. This May, I spied a robin carrying food and watched it until it went into its nest in the bush outside my house. In the nest were five small birds that the parents fed frequently over the next few days. Several years ago, I was watching sandpipers on a beach. If I had simply determined what species they were and looked away I would have missed a special opportunity. I saw the sandpipers freeze and begin looking upward for longer periods of time. By following their gaze, I was lucky enough to see a Peregrine Falcon swooping overhead.

As a professional ornithologist, I use observations to guide the questions that I ask and the data that I collect. During graduate school, I began studying bird migration on an island off the coast of Maine. My initial observations indicated that there were more birds using the island during the spring than during the fall. Because of this difference, I began looking at how the birds were using the site differently between the seasons. My data, gathered through a combination of observation and capture-and-release techniques, showed that birds were more likely to stay on the island for several days and to gain mass during the fall. That observation led me to investigate whether "skinny" birds would be more likely to stay on the island for longer periods of time. Indeed, during the fall "skinny" birds were more likely to stay for several days, but we didn't see that difference during the spring. These observations of variations in bird behavior have led me to investigate other differences between spring and fall migration.

Like most biologists, I collect two types of data: data intended for specific projects and data that may be useful for a future project. For example, several years ago I began collaborating with the Maine Medical Center on a project on Lyme disease. We were interested in whether migratory birds were contributing to the northward expansion of the disease. As part of my studies of songbird migration I collected all the ticks we found on captured migratory birds and sent them to my collaborators, who then identified them and checked them for the bacterium that causes Lyme disease. Our results showed that birds were likely to have transported north ticks carrying Lyme disease. Several years after that study was completed one of my research associates began wondering if the ticks were having an effect on the birds' migration. Using the data we had earlier collected, we found that our most common species of birds were more likely to have ticks in the spring than the fall, but that the number of ticks didn't differ between the seasons. We didn't find any negative effects of the ticks on the migrant birds—the body mass, recapture rate, and length of stopover were similar between birds with and without ticks.

There are still many unanswered questions about the biology and behavior of most species of birds, even the common ones. For example, although we know that American Robins generally have two nests each season, the amount of time between the fledging of the first nest (the time the young leave the nest) and the initiation of the next nest (egg laying) is not yet known. There are also many questions about the robin's song—how does it develop, how much variation exists, and what are the functions of different songs? Detailed observations of behavior may help answer some of these questions.

Finding Investigable Questions

Tracing the Development of Questioning Skills

Formulating investigable questions is key to the inquiry process. It is a challenging task, but one that can be guided and successfully developed in young learners. Without some prior knowledge of the study site and some skill at recognizing which types of questions lend themselves to firsthand investigation, students find it a daunting task to come up with their own questions. So it is very useful to begin by providing some preliminary, guided investigations where an overarching question is posed to the whole class. As a side benefit, it is also easier for the teacher to manage an investigation when the whole class is focused on the same question.

In the beginning stages, students are acquiring a multitude of skills and are asked to assimilate them at a rapid pace. In Chapter 3, Getting Started with Students, they tackle making close observations of preselected objects, and record these in writing and by drawing in their journals. They begin to establish criteria for good journal entries, and one of these criteria includes keeping a log of questions. They look at their first questions and then briefly classify them into those that could be answered by further observation and those that would require using other resources. The point is made that there are indeed different types of questions, and some are more fruitful than others, but all are important and valid products in the inquiry process.

Questions and the Inquiry Standards

In *Inquiry and the National Science Education Standards: A Guide for Teaching and Learning* (National Research Council 2000, 169), different types of questions are given a prominent place, and are described as follows:

Different kinds of questions suggest different kinds of scientific investigations. Some investigations involve observing and describing objects, organisms, or events; some involve collecting specimens; some involve experiments; some involve seeking more information; some involve discovery of new objects and phenomena; and some involve making models.

In Chapter 4, a broad, overarching question focuses the whole class in a journaling activity. The question is provided, and students use it to discover what is at the site, what resources they have to work with. Again, students are grappling with a multitude of newly acquired skills, and in addition, are putting them into practice in a new and unpredictable environment, the outdoors. They have little formal prior knowledge of the study site (even though it may be their own schoolyard), and are concentrated on building a database from which new questions will eventually emerge. Now students' recorded questions begin to take on more importance, for they are stockpiling a series of avenues that may lead them to further exploration.

In Chapter 5, Mapping the Site, the broad question about what is at the site is again implied, but students define it more precisely as they work to record the locations of objects on a map. In addition to cataloguing the resources at the site, students are now also recording the spatial relationships of features and organisms they find there. This is rich material for developing questions, and their stockpile should grow immensely. By now students are actively wondering about the reasons why some organisms are abundant in one place but not another, how one organism influences another, what causes temperatures to differ from place to place, or the relationships between plants, animals, and human activity at the site.

Thus, students are gradually and purposefully prepared to launch more self-directed investigations that pertain to their own questions. They have acquired an immediate and firsthand familiarity with the site, they have applied and practiced important inquiry skills, and they have begun to wonder. Their own questions are now paramount. They have a personal investment in their questions, and are motivated to find their own answers. Moreover, they have developed many of the tools they need to work productively and more independently.

Chapters 6, 7, and 8 are all conducive to this more advanced level of individual or small-team investigation, and all involve a commitment to a more long-term, ongoing study. Students also make real use of data they have already collected about the site, and in most cases will find that they need to expand their databases to incorporate more detail in order to answer a question.

Both in Chapter 6, Creating a Field Guide, and in Chapter 7, Making a Collection, students are encouraged to pursue questions about particular organisms that are of special interest to them. They review their storehouse of ideas and questions, and with careful guidance from the teacher, decide upon one that is investigable, and plan their own investigation.

Student questions relating to identification and classification, habitat, range, and distinctive features of an organism lend themselves well to the creation of a field guide as an organizing tool for the data. The same questions might also be used as the basis for making a collection. In addition, collections can be used to trace an organism's life cycle over time, the distribution of an organism at a site, or variations within a species.

In the process of creating a field guide or a collection, students become "experts" on their chosen organism. But these two vehicles also have more wide-ranging implications because they represent a more formal communication of a question and its answers to a wider audience. Students are no longer just keeping a personal journal account, but are attempting to educate others about the site. Field guides and collections are concrete and enduring pictures of a site at a particular moment in time, and as such, provide others with the natural history of that location at that time.

Many students will have generated questions about animal behavior, and these, being the most challenging but perhaps the most intriguing, are reserved for Chapter 8. Before attempting an animal behavior study, students need to learn additional, rather sophisticated skills. They will first review their questions and try to distinguish between those that deal with animal actions (behaviors) as opposed to physical characteristics; and between observable behaviors as opposed to interpretations of those behaviors (anthropomorphism). In other words, it's not about why an animal is doing something, or how it looks, or what it is feeling; it's about what the animal is doing, and the questions students generate need to reflect exactly that.

Once students have mastered the new ideas and formulated a preliminary question, they will need to spend a good deal of time just observing the animal, listing its discrete behaviors, generating more questions, and looking for behavioral patterns. Finally, they are ready to refine their questions and select what they consider to be an investigable question for in-depth study.

Where Do Students Go from Here?

Guided experiences have paved the way for students to become more independent inquirers. As with good parenting, the goal is to give our children a solid upbringing and then let them go to find their own way. In scientific inquiry, that upbringing includes providing students with a full range of skills, developing in them a heightened interest in the world and its phenomena, expanding their awareness of the possibilities for scientific research, and carefully nurturing their escalating levels of curiosity.

In the guided inquiry mode, as depicted in the early chapters of the book, the teacher takes most of the responsibility for creating the questions for investigation, and shepherds students through the steps of planning and designing their studies. As students move toward more open inquiry, they assume

Essential Feature	Variations			
1. Learner engages in scientifically oriented questions	Learner poses a question	Learner selects among questions, poses new questions	Learner sharpens or clarifies question provided by teacher, materials, or other source	Learner engages in question provided by teacher, materials, or other source
2. Learner gives priority to **evidence** in responding to questions	Learner determines what constitutes evidence and collects it	Learner directed to collect certain data	Learner given data and asked to analyze	Learner given data and told how to analyze
3. Learner formulates **explanations** from evidence	Learner formulates explanation after summarizing evidence	Learner guided in process of formulating explanations from evidence	Learner given possible ways to use evidence to formulate explanation	Learner provided with evidence
4. Learner connects explanations to scientific knowledge	Learner independently examines other resources and forms the links to explanations	Learner directed toward areas and sources of scientific knowledge	Learner given possible connections	
5. Learner communicates and justifies explanations	Learner forms reasonable and logical argument to communicate explanations	Learner coached in development of communication	Learner provided broad guidelines to use to sharpen communication	Learner given steps and procedures for communication

More _____ **Amount of Learner Self-Direction** _____ Less

Less _____ **Amount of Direction from Teacher or Material** _____ More

From *Inquiry and the National Science Education Standards* (National Research Council 2000, 29)

FIGURE 9–1. Essential features of classroom inquiry and their variations

more and more of that responsibility. As described in *Inquiry and the National Science Education Standards* (Figure 9–1), students may need to begin with questions provided by the teacher in order to become more and more proficient at recognizing good questions, and at sharpening or clarifying questions. They may also need to practice by learning to select among questions provided. The ultimate goal, however, is that students pose new, investigable questions of their own and devise their own plans for seeking answers with minimal input from the teacher.

What Is an Investigable Question?

In natural history studies such as those featured in this book, investigations are most often based on observation rather than experimentation. In this case, questions arise directly from observations of the objects or the phenomena (Figure 9–2). Students then gather evidence and data through more and more detailed observations that include using the senses, measuring and quantifying, and repeatedly monitoring the study subjects over time. In drawing conclusions or offering explanations, students make use of their own data and construct their own understanding based on that data. When a study includes all of these elements and therefore requires students to take direct action, we can say that it has been based on an investigable question.

Indeed, formulating an investigable question is the first part of the research process, whether the scientist is in elementary school, graduate school, or working in a commercial laboratory. At all levels, the scientist must also consider if the question is worthwhile, doable in the time and with the materials available, interesting, and practical.

FIGURE 9–2. Firsthand observations can lead to new questions for investigation.

Changing Noninvestigable Questions

Questions that are noninvestigable (at the pre-college-student level) usually have an inherent problem. Students may ask why something is so, for example, "Why does a skunk stink?" or "Why do insects have six legs?" These so-called *why* questions are beyond the range of what students can reasonably tackle.

But *why questions* may sometimes be restated or slanted in an investigable direction. Instead of *why*, the student may ask *how*, and thus be able to observe the phenomenon firsthand. If you asked instead, "How many different ways does a cricket use its legs?" or "How does a ladybug walk? How many legs are on the ground at a time when a ladybug is walking?" there is a good possibility that you could find out yourself through direct observation.

The skunk question is noninvestigable for other, obvious reasons. It would be nearly impossible to observe the subject safely, and even if you did, observation wouldn't answer the *why* question. If you changed the question to something like, "What is in skunk spray that makes it stink?" then you would probably have to consult secondhand resources to find an answer or somehow obtain a spray sample and analyze it chemically using special equipment.

Other problems inherent in a question may be of a more practical nature. For students at the middle school level, for example, overly ambitious questions requiring sophisticated materials or exotic locations are very appealing. But unless students can really have access to the heat shield from the lunar module, for instance, or gorillas in their natural habitat, even exciting investigable questions have to be given up in favor of those that may be more pedestrian but more feasible.

Guided Development

The progression involved in developing the ability to formulate an investigable question is deliberate and carefully choreographed, and it takes time. Students benefit from practicing first in guided situations where they acquire background information and prerequisite skills, but where they continue to keep a record of where they want to go next because of the new, compelling questions that emerge as they work. Students also benefit from exercises in which they analyze types of questions and learn to identify those that they can answer by taking action on their own using the resources available. And finally, even if the question seems wonderful at the outset, students learn that sometimes they need to revise and refine it in order to make it work.

10

Why Teach Inquiry-Based Natural History Science?

How Is Inquiry-Based Natural History Science Different from Experiment-Based Lab Sciences?

In the introduction, we talked about "teasing science out of natural history" and acknowledged that the study of natural history is different from that of experiment-based sciences. Yet despite the differences, the elements of true science are there in equal measure for both disciplines.

Natural history study begins with the most fundamental of scientific skills, observation. In *Outdoor Inquiries*, observations focus on commonplace objects and phenomena in the immediate environment. The objects or phenomena provide the real impetus to collect data. They stimulate questions, pique curiosity about what this object might be, where it came from, why it is there, and what relationship it has to other objects in the area. The object then becomes the catalyst for moving outdoors to find out more about the complex environmental systems that make up the local habitat.

In natural history studies, the observation period may be quite prolonged, for it takes time to notice, to wonder, and to discern patterns. It may also take considerable time before investigable questions arise from the observations. This stage requires patience and persistence, two scientific habits of mind that we would hope to nurture in our students.

From repeated observation eventually comes the need to keep track of what we are noticing, and authentic data collection naturally follows. Data collection may take many forms, as we have seen: keeping a notebook or journal, mapping a site, making a collection, creating a field guide to the local environment, or developing ethograms to track animal behavior. And thus, a scientific investigation is set in motion.

Natural scientists plan their investigations. It is an orderly process. They predict, develop hypotheses, collect and record data, interpret data to draw conclusions, communicate findings, and reflect on the content and the processes in the same way that experimental lab-based scientists do. But unlike lab-based scientists, natural scientists are concerned with deciphering the mysteries of nature's experiments as they already exist in the real world. They work with objects and phenomena as they find them, and attempt to understand what is, but not necessarily what can be manipulated or tested under artificial conditions.

Experimental science and natural history science are not at odds, but share many of the same fundamental principles. Both these branches of science demand a kind of rigor that includes scrupulous attention to detail, honest and accurate record keeping, interpretations based on data, and a straightforward communication of results.

The Inquiry Continuum

Inquiry-based science comes in many different varieties, and may be thought of as a continuous sequence in which many of the stages overlap, but where the two extremes are quite distinct. Guided inquiry in which the teacher plays a strong role belongs to the beginning, fundamental stages while open inquiry in which the student assumes much of the responsibility for learning is at the other end of the spectrum. Students with the least experience with inquiry will need the most structure, and as they become more adept, will gradually be able to take steps toward more self-directed independent work.

Although it is difficult to place *Outdoor Inquiries* squarely in one part of the continuum, it is safe to say that we have concentrated on laying the foundations. We have attempted to provide an array of structured activities to build the skills necessary to do inquiry as well as to suggest springboards, ways to take inquiry further.

The five database strategies focus on the building blocks of inquiry. Students engage in long-term close observation and record these in writing and by drawing. They collect data in five different formats. The database then becomes the evidence for analyzing results and for drawing conclusions. Since formulating investigable questions is a crucial inquiry skill, students have considerable practice in identifying types of questions and make a start at developing their own questions.

In moving toward more open inquiry, students would need to hone questioning skills further. Ideally, students would then design and implement their own scientific investigations, making plans, using appropriate equipment, collecting data over time. They would work to use the evidence to support their claims, and thus refine their analytic skills. Students would take the lead in finding other resources to bolster their arguments, and would communicate their findings logically and in greater detail—all with minimal input from the teacher.

We encourage you to continue leading students toward the open inquiry end of the spectrum. It is a powerful way to become a true learner.

The Real Thing

The projects and activities outlined in *Outdoor Inquiries* integrate with many other disciplines in the curriculum in a natural, authentic way. Since the activities themselves are reality based, the skills that students apply to them are used in a real-world context.

The objects and phenomena in the natural world provide the stimuli to spur discussions and debate among students. Peer teaching becomes a genuine and regularly occurring event as students become "experts" on their chosen topic and communicate their findings, voice their opinions, and pose new questions. The dialogue is rich, and much of the learning arises not from texts or direct teaching, but from shared experiences.

In addition to the obvious links to language arts, math and measuring, social studies and history, art, and technology, there is another less tangible link that is difficult to describe and impossible to teach. It is the personal link that each student makes to the natural world through direct experience. It is not just a cerebral experience that we are hoping for, but an emotional one as well.

Students may express a wide range of responses to their first outdoor field trips. Many will approach with enthusiasm, but for those with little outdoor experience, the unknown and unpredictable situation may summon up only indifference, or possibly even provoke anxiety. These students may initially be overwhelmed by the sheer number of unrecognizable phenomena they encounter outdoors. Or they may have come with negative preconceptions about insects, poisonous plants, or dirt. All of these, while being legitimate emotional responses, not only hinder learning, but also present barriers to building a healthy environmental awareness.

Gradually, and with repeated positive experiences, students can and will become more and more comfortable as they become more competent working in the outdoor setting. Familiarity in this case breeds confidence. And confidence allows enthusiasm to flourish. Countering fears and negativity with positive exposure to the real world can lead to a strong personal bond with nature that enriches a student for a lifetime. For a student who has built a personal database of knowledge about one small corner of the natural world is better prepared to grow in enthusiasm, sensitivity, and appreciation for all its wonders.

Appendix A

Alignment with Standards: Outdoor Inquiries *and the National* Science Education Standards

The recommendations described in *Outdoor Inquiries* are based on the best practices and examples contained in the *National Science Education Standards* (National Research Council 1996).

> The physical environment in and around the school can be used as a living laboratory for the study of natural phenomena. Whether the school is located in a densely populated urban area, a sprawling suburb, a small town, or a rural area, the environment can and should be used as a resource for science study. (45)

> Specific connections between the National Science Education Standards (NSES) and *Outdoor Inquiries* are outlined in the table on page 118.

Guide to the NSES Content Standard A: Science as Inquiry (Chapter 6)	Corresponding chapters of *Outdoor Inquiries* that enable the development of the abilities and understandings described in the NSES
Abilities Necessary to Do Scientific Inquiry	
• Ask a question about objects, organisms, and events in the environment. Identify questions that can be answered through scientific investigations.	3, 4, 5, 6, 7, 8
• Employ simple equipment and tools to gather data and extend the senses. Use appropriate tools and techniques to gather, analyze, and interpret data.	3, 4, 5, 6, 7, 8
• Use data to construct a reasonable explanation.	3, 4, 5, 6, 7, 8
• Communicate investigations and explanations. Communicate scientific procedures and explanations.	3, 4, 5, 6, 7, 8
• Design and conduct a scientific investigation.	6, 7, 8
• Recognize and analyze alternative explanations and predictions.	
Understandings About Scientific Inquiry	
• Scientific investigations involve asking and answering a question and comparing the answer with what scientists already know about the world.	3, 4, 5, 6, 7, 8, 9, 10
• Types of investigations include describing objects, events, and organisms; classifying them; and doing a fair test (experimenting).	3, 4, 5, 6, 7, 8
• Simple instruments, such as magnifiers, thermometers, and rulers, provide more information than scientists obtain using only their senses.	3, 4, 5, 6, 7, 8
• Good explanations are based on evidence from investigations.	3, 4, 5, 6, 7, 8
• Scientists make the results of their investigations public; they describe the investigations in ways that enable others to repeat the investigations.	3, 4, 5, 6, 7, 8
• Scientists review and ask questions about the results of other scientists' work.	3, 4, 5, 6, 7, 8

Appendix B

Resources and References

Vendors

Many of the materials you will need for *Outdoor Inquiries* are readily obtainable at hardware stores, art supply stores, garden shops, discount stores, and drugstores. The list below provides suggestions for vendors who specialize in science education products not commonly available.

American Science and Surplus: www.sciplus.com

Carolina Biological Supply Co.: www.carolina.com

Classroom Direct: www.classroomdirect.com (rulers, paintbrushes)

Museum Products: www.museumproducts.net/museum/ (hand lenses)

Nasco International/Educational Materials: www.enasco.com (forceps, thermometers, nets, pH testing kits)

Nature Study Guides: http://home.att.net/~naturebooks/index.html (field guides)

Science Kit and Boreal Laboratories: www.sciencekit.com

United States Plastics Corp.: www.usplastics.com (collection jars and bags)

Websites

For an up-to-date directory of useful websites, also see the list we provide online at www.firsthandlearning.org.

Discover Life: www.discoverlife.org

This website provides online field guide databases for help with identification of plants and animals.

National Wildlife Federation: www.nwf.org

This high-quality website has online field guides for both plants and animals. It also offers guidance in developing schoolyard habitats.

Roger Tory Peterson Institute of Natural History: www.rtpi.org/eduprogs

A nationally recognized organization based in Jamestown, New York, RTPI offers outstanding outdoor education programs. In addition, RTPI produces www.enaturalist.org, an online program providing weekly environmental education units. Each unit has artwork, text, activities, and additional websites, plus online access to a professional naturalist.

References

The works cited below are an eclectic mix of resources. They include field guides, how-to manuals, in-depth natural history references, and personal memoirs.

Bourne, Barbara, ed. 2000. *Taking Inquiry Outdoors: Reading, Writing, and Science Beyond the Classroom Walls.* **York, ME: Stenhouse.**

The editor describes this resource as "a book of reflections on children and learning; on teaching; on science made understandable through reading, writing and hands-on investigations—all within the context of the outdoors."

Campbell, Brian, and Lori Fulton. 2003. *Science Notebooks: Writing About Inquiry.* **Portsmouth, NH: Heinemann.**

The authors tell how to integrate writing into an inquiry-based science program, and provide many examples of notebooks.

Dillard, Annie. 1987. *An American Childhood.* **New York: Harper and Row.**

As a child in Pittsburgh, Dillard finds natural history subjects in alleyways, in books, and in parks.

Douglas, Rowena, ed. 2006. *Linking Science and Literacy in the K–8 Classroom.* **Arlington, VA: NSTA Press.**

Researchers and professional development experts offer a broad range of perspectives on the science and literacy connections.

Grant, Tim, and Gail Littlejohn, eds. 2001. *Greening School Grounds: Creating Habitats for Learning.* **Toronto: Green Teacher.**

An anthology from *Green Teacher* magazine, giving step-by-step instructions for numerous schoolyard projects, from tree nurseries to school composting to native-plant gardens, along with ideas for addressing the diverse needs of students.

Heinrich, Bernd. 1989. *Ravens in Winter.* **New York: Summit Books.**

Vermont biologist Heinrich carries out a methodical investigation of the scientific literature before setting out to answer the question of whether ravens recruit each other to food caches in winter. His tale of experiment and observation grows out of detailed field notes and drawings. It repeatedly demonstrates how much there is to learn from failed hypotheses.

Heinrich, Bernd. 1997. *The Trees in My Forest.* **New York: Cliff Street Books (an imprint of HarperCollins).**

The author paints a scientific yet sometimes poetic picture of his own three-hundred-acre forest in all its complex relationships.

Leslie, Clare Walker, and Charles E. Roth. 1998. *Nature Journaling: Learning to Observe and Connect with the World Around You.* **Pownal, VT: Storey Books.**

This book offers simple techniques that will help you capture in sketches and words your ongoing experiences with nature. Readers will find exercises that will develop their observation skills and help overcome their fear of drawing.

Long, Kim. 1995. *Squirrels: A Wildlife Handbook.* **Boulder, CO: Johnson Books.**

This illustrated handbook offers concise, accurate facts about squirrels, their habitat, behaviors, and relationship to man.

Lowenstein, Frank, and Sheryl Lechner. 1999. *Bugs: Insects, Spiders, Centipedes, Millipedes, and Other Closely Related Arthropods.* **New York: Black Dog & Leventhal.**

This large-sized book is loaded with full-color photos that magnify the amazing physical features and behaviors of dozens of insects and other arthropods. While the text is substantial, the exceptional photos will engage all ages.

Mason, Rich. *Schoolyard Habitat Project Guide.* **U.S. Fish and Wildlife Service. http://chesapeakebay.fws.gov/pdf/habitatguide.pdf.**

This is a comprehensive and very informative guide to developing schoolyard habitats.

Moline, Steve. 1995. *I See What You Mean.* **Portland, ME: Stenhouse.**

An activity book that shows how to help students communicate graphically with over one hundred examples of student work.

National Research Council. 1996. *The National Science Education Standards.* **Washington, DC: National Academy Press.**

———. 2000. *Inquiry and the National Science Education Standards: A Guide for Teaching and Learning.* **Washington, DC: National Academy Press.**

Peterson, Roger Tory. The Peterson Field Guides series. Boston: Houghton Mifflin.

There are nearly fifty different Peterson Field Guides. Each features detailed descriptions, color photos, black and white drawings, and distribution maps for a variety of plant or animal species and subspecies. Information on how to use the guides is included.

Rezendes, Paul. 1995. *Tracking and the Art of Seeing: How to Read Animal Tracks and Sign.* Charlotte, VT: Camden House.

Have you ever encountered a track, a nibbled branch, or a hole in the ground and asked, "What kind of animal made that?" Whether novice or expert, this rich resource will help you pursue and answer that question as well as many other questions about wildlife.

Russell, H. R. 1997. *Ten-Minute Field Trips: Using the School Grounds for Environmental Studies.* Washington, DC: National Science Teachers Association.

A collection of mini–field trip ideas, the book contains over two hundred ways for children to investigate the local environment.

Saul, E. Wendy, ed. 2004. *Crossing Borders in Literacy and Science Instruction.* Arlington, VA: NSTA Press.

Two dozen essays by well-known educators challenge teachers to develop the interdisciplinary links between science and literacy.

Saul, Wendy, and Jeanne Reardon, eds. 1998. *Beyond the Science Kit: Inquiry in Action.* Portsmouth, NH: Heinemann.

Essays written by teachers describe how to go beyond prescribed science kits to authentic scientific inquiry.

Schwartz, Charles Walsh. 1981. *The Wild Mammals of Missouri.* University of Missouri Press.

This comprehensive book offers informative portraits of dozens of mammals, including illustrations and photographs of their skulls, scat, tracks, and habitats.

Stokes, Donald, and Lillian Stokes. Stokes Nature Guides. Boston: Little, Brown.

This series of wonderful guides emphasizes animal behavior, ecology, habitat, and life cycle.

Symonds, George W. D. 1958. *The Tree Identification Book.* New York: Quill.

Discover a convenient method for recognizing and identifying trees using keys designed for easy visual comparison. This guide includes many photographs that illustrate in detail a wide variety of leaves, fruit, bark, twigs, and buds.

Whitin, Phyllis, and David J. Whitin. 1997. *Inquiry at the Window: Pursuing the Wonders of Learners*. Portsmouth, NH: Heinemann.

A yearlong study of birds presents opportunities for students to learn through inquiry.

Wilson, E. O. 1994. *Naturalist*. Washington, DC: Island Press.

In this engaging memoir, Wilson describes how his firsthand experiences with nature as a boy set him on a path to becoming one of the greatest natural scientist of our times.

Periodicals

Connect. PO Box 60, Brattleboro, VT 05302-0060, www.synergylearning.org.

Connect is published bimonthly during the school year and offers a wide range of practical, teacher-written articles. Each issue is thematic and supports hands-on learning, problem solving, and multidisciplinary approaches. Published by Synergy Learning, a not-for-profit corporation engaged in publishing and professional development for pre-K to middle school educators.

Green Teacher. PO Box 452, Niagara Falls, NY 14304-0452, www.greenteacher.com.

Published four times a year, *Green Teacher* is a magazine by and for educators to enhance environmental and global education across the curriculum at all grade levels.

Science News. PO Box 1925, Marion, OH 43306, www.sciencenews.org.

Science Service, a nonprofit corporation, publishes this weekly news magazine. It contains short, well-written articles about current developments in all fields of science and technology. It also includes short reports on recently published books.

Science Scope. NSTA, 1840 Wilson Boulevard, Arlington, VA 22201-3000, www.nsta.org.

This is the National Science Teachers Association's monthly journal for middle school teachers. It contains well-written articles by teachers and other educational professionals on topics relevant to middle school teaching. It is available with membership in the NSTA.